T0077974

An Assessment of the Probability of the Existence of Extraterrestrial Technological Civilizations or

IS ANYBODY OUT THERE?

Second Edition

W.H. Collier

authorHOUSE®

AuthorHouse™
1663 Liberty Drive
Bloomington, IN 47403
www.authorhouse.com
Phone: 833-262-8899

Published by AuthorHouse 04/21/2022

ISBN: 978-1-6655-5777-1 (sc)
ISBN: 978-1-6655-5778-8 (e)

Library of Congress Control Number: 2020919099

Print information available on the last page.

Editor: Lauren Hendrick
Associate Editor: Kerri Davidson
Production Manager: Adrienne Metz
Cover Design: Margaret Girouard

This book is printed on acid-free paper.

ACKNOWLEDGEMENTS

My good friend, Harley McElroy, asked me this very same question at a dinner party, to which I said I would have to get back with him. This Assessment twenty years later is my response.

Many thanks to W.C. Atkinson, who consulted with me on this project on a hundred or so Sundays (for the modest fee of a hundred or so cocktails).

Thanks as well to my wife, Nicole, who, if never completely confident in this project, was always completely confident in me, which is what really counts.

CONTENTS

THE MISSION

For starters, I want to make one thing perfectly clear. There is no evidence whatsoever that spacemen have ever visited this planet. Nor is there a shred of evidence that flying saucers or other alien spacecraft have ever entered Earth's airspace. Further, neither NASA nor the ESA nor SETI nor anyone else has ever detected a transmission from an intelligent, extraterrestrial source. The myths surrounding Roswell, Area 51, crop circles, Nazca lines and the Easter Island Moai are just that. As far as we know, there is no life outside this planet. We are looking but have found nothing yet. Absolutely nothing. So, with that caveat securely in mind, let us proceed.

Of course, just because we have found no evidence of extraterrestrial life does not mean there is none. The absence of evidence, as scientists are fond of saying, is not the evidence of absence (though it does mean that the matter remains unproven). Besides, until recently, we have not been looking very hard or very effectively.

The folks at the SETI Institute (i.e., Search for Extra Terrestrial Intelligence) have been, since the early 1980s, scouring the heavens in the search for evidence of intelligent life by looking for the signatures of its technology (radio and other electromagnetic signals, that is). So far, they have found nothing.

During the last couple of decades, however, the search for extraterrestrials has focused on the search for exoplanets (that is, planets outside of our solar system). This has proven much more successful. The first discoveries of exoplanets were made by Earth-based telescopes. More recently, though, the major discoveries have been by space-based telescopes.

1

By making observations from above the Earth's atmosphere, astronomers enjoy clarity of view and are able to detect wavelengths across the entire electromagnetic spectrum.

Designed to study the cosmos at infrared wavelengths, the Spitzer Space Telescope has made significant contributions to the study of exoplanets. Launched in 2003, Spitzer spent sixteen years making observations with its 33-inch primary mirror. The data developed by the telescope has enabled astronomers to determine the temperature and atmospheric structure, composition and dynamics of a number of exoplanets.

The most successful exoplanet finder of them all was NASA's Kepler Space Telescope. Its mission was specifically to survey our region of the galaxy to find exoplanets. And that it did. From the time of its launch in 2009 until its retirement in 2018, Kepler discovered some 2,700 exoplanets using the transit method of detection (which we will discuss later).

TESS (the Transiting Exoplanet Survey Satellite) is another successful exoplanet hunter, if not so spectacular as Kepler. Launched in 2018, it remains in operation to this date. While Kepler observed stars in a single patch of sky, TESS views an area 400 times larger, covering about 85% of the entire sky. To date, TESS has discovered 66 confirmed exoplanets, along with some 2100 unconfirmed candidates.

The European Space Agency's CHaracterizing ExOPlanets Satellite (CHEOPS) was launched in December of 2019. Its mission is not to search for new exoplanets but to study in greater detail several hundred previously discovered exoplanets.

Many more exoplanets will be found by *Big Daddy* of space telescopes, NASA's James Webb Space Telescope. Webb is the successor to the phenomenal Hubble Space Telescope but is far more powerful. While Hubble had a primary mirror 8.0 feet (2.4 meters) across, Webb's is 21.3 feet (6.5 meters). In January of 2022, the satellite arrived at its permanent location a million miles from Earth at a point referred to as L_2 (i.e., Lagrange Point 2, a Lagrange point being a position in space where the gravitational force of two large masses precisely equals the centripetal force required for a small object to move with them. In other words, where an object sent there will stay put). Its first photos are expected by midsummer of the year. Stay tuned.

Of great interest and importance to our inquiry will be the ESA's PLAnetary Transits and Oscillations of stars (PLATO) telescope, planned for launch in 2026. It will search specifically for Earth-sized planets transiting stars similar to our Sun.

As you can see, we are now looking hard, and, with ever improving instruments, more and more effectively. So, we just might have that evidence of extraterrestrial life in the very near future. Or maybe not. We will see.

Despite the present lack of evidence, I think most scientists feel quite certain that life and even intelligent life is abundant in the universe. The reason for their confidence is the staggeringly large number of planets there must certainly be. In our galaxy alone there are 200 billion stars, most of which are expected to have multiple planets. With perhaps a trillion planets, the galaxy must be just teeming with life, or so the reasoning goes, if that can be considered reasoning at all, which, I think, it cannot.

Considering the magnitude of the question, I'd say a little more inquiry is in order. Life requires a planet suitable for life. Most planets are not. In our own solar system, there are eight planets, only one of which can support life. Wouldn't you want to know how many of the trillion or so planets are actually habitable? Can we expect all habitable planets to be inhabited?

3

If not, why not? What are the necessary preconditions for the emergence of life? Is it easy or is it hard? Where there is life, will intelligent life eventually emerge? Is there a trajectory toward intelligence? How did intelligence arise on this planet? Does our experience have any relevance to the rise of life and intelligent life on other planets? These are the kinds of issues that must be addressed before we can make any reasonable assessment of the likelihood of anybody being out there. So, that is precisely what we will attempt to do.

Making such an assessment is a multidisciplinary project, involving astronomy, earth sciences, biology, microbiology, paleontology, paleoanthropology and more. That may be the reason why there have not been more of these analyses. Most scientists are simply not comfortable venturing outside of their own field of expertise. Moreover, to do so might be seen as an unprofessional encroachment on others' turf. I am not a scientist, so I have no such compunction. I am an historian, writer and attorney. Although I have read widely on these topics over the past couple of decades, I have no formal background in science. That would seem to make me singularly unqualified to undertake the present investigation. My job, however, is not to offer my own insights and opinions but rather to collect and present the ideas of the leading experts in each field and compile the latest scientific consensus on each topic. That, I am qualified to do.

As the title says, the goal of this work is to make a reasonable assessment of the probability of the existence of extraterrestrial technological civilizations, meaning extraterrestrial civilizations with the capacity to communicate with civilizations on other planets also having that capacity. To do so, we will, in Part I, first attempt to determine what conditions are necessary for a planet to be considered habitable and how many habitable planets there might be. In Part II, we will examine what is necessary for life to emerge and develop on one of these habitable planets. Finally, in Part III, we will see what it takes for an intelligent species to arise on one of our living worlds.

Although I am not a scientist, I was, at one nadir of my life, a trial attorney, which gave me a keen appreciation of the difference between actual evidence and unsubstantiated assertions. In this assessment, I will stick to the evidence, at least where available, and limit speculation to those areas where conjecture is unavoidable.

This, of course, is not the first such analysis. In 1961, the astronomer, Dr. Frank Drake, composed his famous Drake Equation to estimate the number of communicating civilizations in the galaxy. Drake concluded that the chances that we have been the only technologically advanced life form to have ever developed is about one in a gazillion (he actually said one in ten billion trillion, but I think that's about the same thing). Consequently, he opined, the chances that another advanced civilization has evolved is "astonishingly likely." I agree with some of the factors used by Dr. Drake in his equation and will consider them in this inquiry, but much has been learned in the more than half a century since he performed his calculations. What is needed now is a fundamental reassessment of the probabilities of extraterrestrial life based on a comprehensive review of the very latest astronomical, astrophysical and astrobiological data, using the most sophisticated scientific methodologies and subjecting the findings to a rigorous, scholarly, peer reviewed analysis. Unfortunately, this is not that. This will be only a gathering of what is presently known and an attempt to draw from the data logical conclusions. It will, however, be conducted dispassionately and objectively, allowing the evidence to lead where it will.

And so, with the recognition that a little knowledge can be a dangerous thing, let's get to work.

I

HABITAT

If a technologically advanced civilization is to exist, it must have a home, or, at least, it must have had a home from which it arose. So, our initial challenge will be to quantify the number of planets there might be that would be suitable for spawning, nurturing and developing life.

By "suitable for life" I mean suitable for carbon-based life like that on our planet. There is good reason to so limit our inquiry. Life is chemistry and, as such, requires a vast number of chemical compounds to build structures, to produce, store and use energy, to manage and regulate chemical processes and to reproduce. The number and variety of biochemical compounds needed for life is staggering. Carbon is the only element that can form the myriad of chemical compounds required for even the simplest life. Why? Basic chemistry. I know we all remember our high school chemistry, but let's review anyway.

Atoms are composed of a nucleus, containing positively charged protons and neutral neutrons, surrounded by a buzzing cloud of negatively charged electrons. To remain electrically neutral, atoms usually have the same number of protons and electrons.

Electrons do not revolve around the nucleus in individual orbits as do planets around the sun. Instead, they are confined within discrete, concentric energy shells surrounding the nucleus. The first and most closely positioned shell can hold up to two electrons. The atoms of the simplest elements, hydrogen and helium, are composed, respectively, of one proton and one electron, and two protons, two neutrons and two electrons.

Helium, with its two electrons, has a fully filled electron shell. Consequently, it is sated and content to just stand around by itself doing nothing at all. Hydrogen, on the other hand, has a burning desire to find another electron to fill its shell. The easiest way to do that is to partner up with another atom that is trying to do the same. So, a hydrogen atom might join together or bond with another hydrogen atom, each sharing its lone electron with the other, effectively filling the shells of both and forming a molecule of hydrogen (H_2). In other instances, multiple hydrogen atoms might enter into a polygamous relationship with an atom of another element altogether, as in H_2O or NH_3, forming the compound substances, water and ammonia, each having entirely different chemical properties from the atoms from which it is formed. That's what we are looking for, chemical reactions that create compound molecules.

For larger atoms, another electron shell is needed. This second shell will hold up to eight electrons. Thus, there are eight second shell elements, each having two electrons in its inner shell and between one and eight electrons in its outer shell. Larger atoms will have additional shells. It is in an atom's last shell, referred to as the valence shell, where all the all the action is. As with hydrogen in the first shell, atoms of the second shell elements having fewer than eight "valence" electrons will react (partner up) with other atoms to fill their respective shells, creating compound molecules in the process.

The largest of the second shell elements is neon, which has ten protons and a corresponding ten elections (two in the first shell and eight in the second). With both of its electron shells full to the brim, neon, like helium, is inert, unreactive and of no use for present purposes. The other seven, however, are reactive, meaning they can join with other atoms to fill up their second shell and create compound substances.

The remaining seven elements have between one and seven valence electrons. The desire of every atom, however, is to have a full outer shell. One way of doing that is to give away electrons. Lithium, for instance, has only one electron in its second shell. Rather than going to the trouble of acquiring seven more electrons, it usually just gives its one away, becoming a positively charged version of itself (i.e., three protons, but only two electrons). The same is true for, beryllium and boron, with only two and three electrons, respectively. Like Lithium, they have only a weak hold on their valence electrons and have no problem letting them loose and existing in a positively charged state.

Conversely, fluorine, with its seven second shell electrons, needs just one more electron to fill out its outer shell and will stop at nothing to get it, typically yanking an electron from an innocent bystander. This ruthless reactivity, however, is not conducive to forming the large stable molecules needed for life. Nitrogen with five valence electrons and oxygen with six have their own problems. Nitrogen atoms tend to bond with other nitrogen atoms in pairs to form the molecule N_2. This molecular nitrogen has a triple covalent bond which requires substantial energy to break apart before the nitrogen atoms can react with other atoms. This limits its reactivity, rendering it essentially inert under most circumstances. Oxygen is finicky and will bond with only the right type of elements. Moreover, with only one, two and three openings, the bonding opportunities for these three elements are limited.

Carbon has four valence electrons and four empty spaces. That's too many electrons to give away and too many spaces to fill by snatching electrons from others. So, carbon atoms must get down to the business of establishing relationships if it is to fill its second shell. With four spaces available, carbon can bond directly with many different elements. Moreover, its bonds are neither so frail as to beak easily nor too rigid to deter adaptability. It can form single bonds or stronger double and even triple bonds. It can also bond with other carbon atoms. Where a pair of carbon atoms forms a single bond, the pair will still have between them six open bonding positions (three for each atom). Add another, and they'll have nine. And on it goes. This allows for the creation of long strings of carbon and other elements, sometimes hundreds of atoms long. The strings can grow linearly with branches that form large networks, or close in to form rings and hexagons that can combine with other structures to form the large, complex macromolecules that are essential to life. The possibilities are almost limitless. Indeed, more than ten million carbon-based compounds are found within living things.

There is, of course, a third energy shell beyond the second. It too holds up to eight electrons. For many of the same reasons discussed above with reference to the second shell, none the third shell elements are viable candidates for serving as a basis of life, except, perhaps, for one. Silicon. Like carbon, silicon has four out of a possible eight valence electrons. So, wouldn't that make it similarly reactive and thus able to perform the sorts of chemistry that carbon does? No.

The third electron shell, you see, is much farther away from the nucleus than the second, and, therefore, its electrons feel a weaker pull of the nucleus. Its small size and the close proximity of its nucleus to its valence electrons allow carbon to form strong bonds with other atoms, creating highly stable compounds. Because of silicon's greater size, the bonds it forms with other atoms are only about half as strong as those formed by carbon. Consequently, silicon cannot form long chain compounds as can carbon. Moreover, carbon bonds strongly with other carbon atoms allowing it to form long chains of carbon atoms, while silicon does not bond well with itself.

Moreover, silicon-silicon bonds break apart in water; silicon bonds weakly with carbon and hydrogen; and silicon bonds readily and very strongly with oxygen, inhibiting its participation in the types of chemical reactions commonly associated with life. There are other reasons as well, but I think that should be enough to safely relegate silicon-based lifeforms to the realm of science fiction.

As we move to even more distant electron shells, the elements grow ever larger and more cumbersome and ever less likely to serve as a basis for the chemistry of life. So, we will pretermit any further discussion of these larger elements.

Since carbon is the only element that can perform the complex chemistry of life, and since no one is seriously suggesting that there could be any other form of life, I think it reasonable and proper to limit our inquiry to planets amenable to carbon-based life.

Further, we will limit our search to our own galaxy. There are perhaps 200 billion galaxies in our universe. For our purposes, that, is not a manageable number. Besides, our galaxy, though fairly large, is not unusual in any meaningful way. Our findings for this galaxy should be equally applicable to most all others.

Finally, we will analyze planets only. Recent findings suggest that some moons, such as Jupiter's Europa and Saturn's Enceladus, might have vast subsurface oceans of liquid water. The pull of gravity on these moons from their planets and other nearby moons causes tidal heating within these moons sufficient for liquid water to exist, despite surface temperatures of as low as minus 300 F. Where there is water there is the possibility of life. Underwater environments, however, are ill suited for the development of technology. So even if there is life on such moons, it is highly improbable that a technologically advanced civilization could develop beneath an ocean, particularly one encased in miles of ice.

Subject to these limitations, we will search first for suitable stars then for suitable planets orbiting such stars.

1.   The Milieu

The universe (this universe?) came into existence with the so-called Big Bang 13.8 billion years ago. At that moment, all of the energy in the universe (there was no matter yet because it was far too hot) was concentrated into a tiny point smaller than the nucleus of an atom. At the moment of the Big Bang, there was neither space nor time. These came into existence when the microscopic point suddenly and violently expanded.

What was there before the big bang? Well, if the big bang created time, then "before the big bang" has no meaning. Asking into what the universe expanded is equally meaningless, since, if the expansion created space, there was no place to expand into. It just expanded. I know, these are not the easiest concepts to get your mind around.

Anyway, the temperature at the Planck time, that is, the moment in time when the universe was 10 million trillion trillion trillionths of a second old (or, as scientists would say, 10^{-43} seconds, where the 43 represents the number of zeroes following the one, as in 10^2 = one hundred, 10^3 one thousand, 10^6 one million, 10^9 one billion, 10^{12} one trillion, and so forth), was a balmy 100 million trillion trillion degrees (10^{32}) Kelvin (180 million trillion trillion degrees Fahrenheit, if that helps). How do we know? Simple math. Simple, that is, if you have a PhD in physics. In the 1920s, the American astronomer, Edwin Hubble, demonstrated that we live in an expanding universe. Thus, the universe will be larger tomorrow than it is today and larger still the day following. If we run the movie in reverse all the way back to the moment of the big bang, all of the matter and energy in the universe will have been squeezed together into that subatomic point. As you might imagine, the confluence of all that matter and energy will cause temperatures to soar, which physicists can calculate with great precision, or pretty much.

At the age of a hundredth of a billionth of a trillionth of a trillionth of a second (10^{-35} seconds), the universe underwent a burst of expansion known as inflation. During that period, space expanded faster than the speed of light from subatomic size to the size of a golf ball.

As the cosmos expanded, temperatures dropped. By one second after the Big Bang, the universe was cool enough for protons, neutrons and electrons and their antimatter counterparts to condense out of an earlier quark-gluon plasma. Three minutes later, the universe had cooled to a billion degrees Kelvin, allowing the nuclei of the lightest elements to form. In another 380,000 years, the universe had cooled to 3000 k, which meant that the electrons had slowed to the point where they could be captured by the nuclei to form the first atoms. In the still intense heat and radiation, only the very lightest atoms could survive, those being hydrogen, comprising 75% of all matter, helium, 25%, and a smattering of lithium. These percentages continue to hold true today, the smattering now including all the other elements.

At that point, gravity began to take charge. Gravity will cause any spot that is even only slightly denser than its surroundings to attract matter in the vicinity. With the addition of new matter, the area becomes denser yet, intensifying its gravitational pull. Over the course of millions of years, clumps of hydrogen and helium gas attract more and more matter, growing ever larger and emptying out the spaces between them in the process.

As matter accumulated, the clumps compressed, increasing the gravitational pressure within. Eventually, the pressure overwhelmed the electrostatic repulsion of the hydrogen atoms, forcing their nuclei together and igniting a fusion reaction. When the universe was 250 million years old, the first stars were born this way, lighting up the darkness of the cosmos.

Soon after the formation of the first stars (or possibly concurrent with their formation), the first galaxies, including our own, began to form. Initially, they were small and dim, containing relatively few stars. They were, however, loaded with gas that would eventually lead to the birth of many more stars, a process that continues to this time. Small galaxies merged with others to form larger and larger structures. Indeed, our own galaxy, the Milky Way, is thought to have formed from the merger of a hundred or more small galaxies.

Today, the observable universe extends 93 billion light years, and it is continuing to expand. There is an unobservable portion of the universe as well. We don't know much about it, since it is too far away to be observed. Observations from the Sloan Digital Sky Survey and the Planck satellite indicate, however, that it must be at least 250 times the radius of the observable part. That means the unobservable universe is 23 trillion light years in diameter and contains a volume of space 15 million times as large as what we can observe.

Einstein's Theory of Special Relativity establishes that the maximum possible speed is the speed of light (186,000 miles per second). The universe is 13.8 billion years old, yet just the visible portion is 93 billion light years across. Apparently, the expansion of the universe does not observe Einstein's cosmic speed limit.

With the expansion of space, the temperature has continued to drop. Today, it is 2.73 degrees Kelvin (-454 F), that is, 2.73 degrees Celsius above absolute zero, which is the temperature at which no energy from molecular motion (i.e., heat) is available for transfer to other systems.

There are about 200 billion galaxies in the observable universe. A recent reassessment of surveys taken by NASA's Hubble Space Telescope indicates that the number should be more like two trillion galaxies. That, however, reflects the number of galaxies in the early universe. Over time, as we have seen, the myriad of primitive dwarf galaxies have merged into the larger galaxies we now observe.

The Milky Way galaxy is one of about 54 gravitationally bound galaxies known to astronomers as the Local Group. All but three are dwarf galaxies. The two largest, Andromeda and the Milky Way, are presently about 2.5 million light years apart. They are, however, on a collision course and will collide in about 4.5 billion years. Together, the two galaxies contain about 1.2 trillion stars. Galaxies are so empty that no two stars will collide. Not even close (the chances that a star will enter within Neptune's orbit are 1 in 10 million). Gravity, though, will wreak havoc on the trajectory of the stars and planets, hurling many out into the void of space. In the end, they will merge into a single massive galaxy.

2.   Goldilocks

The Milky Way galaxy is a barred spiral galaxy, that is, a spiral galaxy with a central bar-shaped collection of stars. It is composed of between 100 and 400 billion stars, the most usual estimate being 200 billion. We will use that.

The galaxy has a diameter of about 100,000 light years (i.e., the distance that light travels in one year, or about 5.9

trillion miles). A view from above would reveal a central bulge surrounded by four large spiral arms. The arms are contained within the galactic disk, which is only about a thousand light years thick.

The question, then, is whether there are some places in the galaxy that are more conducive to life or is one place about as good as any other.

One thing is for certain. Planets orbiting stars within the central bulge are not going to fare very well. There are about 10 million stars within the central one cubic parsec (3.2 light years) of the galactic center. By way of comparison, there are no stars within one parsec of our Sun and only perhaps seventy-four stars within ten parsecs. As a result, the galactic center is awash with harmful radiation, such as gamma rays, x-rays and cosmic rays, that would destroy any life trying to evolve there.

Moreover, catastrophic events, such as supernovas (exploding stars) and gamma ray bursts (stellar explosions that can release more energy in ten seconds than the Sun will in its entire ten-billion-year life) have the capacity to sterilize a planet. If, for instance, a supernova occurred within 10 parsecs of Earth, high energy protons released from the blast would obliterate the ozone layer, leaving life on the surface unprotected from the Sun's ultraviolet radiation. Such catastrophes may be relatively rare, but, with so many stars crammed together, the chances that a planet in the galactic center would experience one or more such events are so high as to be almost inevitable.

It appears that most, if not all, of the planetary systems are surrounded by a halo of rock, ice and other debris left over from the formation of the star and its planets. In our planetary system, this is known as the Oort cloud, named for the astronomer Jan Oort who first theorized its existence. When the cloud of debris is jiggled a little by, say, a passing star, some of the debris is kicked from its orbit and sent hurtling toward the star (and the planets orbiting relatively nearby). We know these chunks of material as comets. When a comet a few miles in diameter travelling a hundred times faster than a speeding bullet collides with a planet, the consequences for any life there are devastating, as the dinosaurs and the seventy percent of all species on Earth that were wiped out by such a collision sixty-five million years ago would attest (okay, it was probably an asteroid, but the effect would be substantially the same). Comets for us on Earth are rare, but, again, increase the number of nearby passing stars, and they would be commonplace.

So, it would seem the farther away from the galactic center the better. This, however, is not exactly the case. In the beginning, you see, there was essentially only hydrogen and helium, the two simplest atoms (hydrogen consisting of one proton and one electron and helium of two protons, two neutrons and two electrons). A couple of hundred million years after the Big Bang, the universe had cooled sufficiently enough to allow gravity to begin pulling the clouds of hydrogen and helium together. As these clumps of gas became dense and internal pressures increased, they began to heat up. As the collapse continued, temperatures continued to rise. When the cores of these protostars (that is, the innermost part of a star) reached a temperature of 15 million degrees Kelvin, nuclear fusion began, and the first stars were born. Fusion is simply the fusing together of simpler atoms to form a more complex atom. In stars, four atoms of hydrogen fuse together to form one atom of helium. Since four atoms of hydrogen contain more energy than one atom of helium, the excess energy is released as light and heat. This released energy pushes outward against the compression of gravity. Eventually, the star reaches a state of equilibrium, where the compressing force of gravity equals the outward push of fusion. It is then referred to as a main sequence star.

A star can remain in balance for millions or billions of years. Eventually, however, the star will use up all of the available hydrogen in its core, and the fusion of hydrogen into helium stops. Without the outward push of fusion, the helium core begins to contract again, causing it to heat up. The increased temperature heats up a shell of hydrogen surrounding the core until it is hot enough to start a new round of hydrogen fusion. The higher temperatures in the core increase reaction rates, and the star's energy output soars by a factor of a thousand. With stars the size of our Sun, the increased energy causes the outer layers to expand, greatly increasing the stars radius and turning it into a red giant. When this happens to our Sun, the radius of the red giant Sun will extend beyond Earth's orbit. The increased size will cause the overall temperature of the star to cool, but the collapsing core continues to heat. When the collapse has increased the pressure in the core sufficiently, fusion begins again, this time fusing helium atoms into carbon and oxygen. Whereas our Sun fuses hydrogen at a core temperature of 15 million degrees kelvin, helium fusion occurs at temperatures ten times higher. The star will remain in this stage for a few thousand to a billion years. When the helium is exhausted, fusion stops. At that point, the upper layers of the star will be ejected, leaving only the dense, white-hot carbon/ oxygen core, known as a white dwarf star.

How hot is a white dwarf? About 100,000 degrees Kelvin (180,000 F). How dense? Well, a white dwarf will have the mass of the Sun but only the radius of the Earth. As a result, the gravity on the surface of a white dwarf is 350,000 times the gravity of Earth, meaning a 150-pound person on Earth would weigh 50 million pounds. Over tens or even hundreds of billions of years, a white dwarf will gradually cool until it becomes a black dwarf, emitting no energy.

In more massive stars, the collapse begins again after the helium is exhausted, until the pressure and heat increase sufficiently to begin another round of fusion, this time fusing carbon and helium into oxygen and oxygen and helium into neon. And on it goes. Neon fuses with helium to form magnesium, magnesium and helium to silicon, silicon and helium to sulfur and on and on until finally chromium and helium produce iron. Iron fusion, however, actually absorbs rather than produces energy in fusion. Thus, once the core has turned to iron, fusion is at an end. With the cessation of fusion, gravity collapses the star and the iron core heats up. In a split second, the outer layers of the star fall in on the core, causing it to heat to billions of degrees and explode. All heavier elements are created during this explosion. The outer layers of the star are cast off, and the debris scattered out into space. This is a supernova that we referred to earlier. Gravity packs the matter in the collapsed core so tightly that the electrons and protons merge to form neutrons, and hence it becomes a neutron star. If the star is sufficiently massive, the core will collapse even further into a black hole.

Neutron stars are even denser than white dwarfs, having 1.4 times the mass of the sun, but a diameter of only 12.4 miles, and a gravity 2 billion times that of the Earth (a tea spoon of neutron star would weigh a billion tons). A black hole at its center is infinitely dense.

Star formation has continued by this same process throughout the nearly fourteen-billion-year history of the universe. The clouds of hydrogen and helium from which later stars were formed, though, might also include carbon, oxygen, iron and all of the other elements formed in the massive stars and cast out into space by a supernova.

The initial stars could not have had planets, since only hydrogen and helium existed. Later generation stars, however, if formed from clouds seeded with these heavier elements, could. To form planets, though, there must be a certain minimum abundance of these heavier elements in the clouds. This suggests that many generations of stars had to form and evolve before there would have been enough heavier elements to permit habitable planets to form.

For no discernable reason, astronomers refer to all elements heavier than hydrogen and helium as metals and the relative abundance of metals as metallicity. Not surprisingly, planets are more likely to form in areas of high metallicity, and planet formation is rare in areas of low metallicity. Because there are more stars toward the center of the galaxy than on the fringe, the relative abundance of the heavier elements is greater toward the center. As one travels away from the center, metallicity decreases, making planet formation less likely and eventually impossible.

So, it seems there is a "just right" region for life in our galaxy that's not too close to the dangerous galactic center, but not too distant so as to be deprived of planet forming metals. Astronomers estimate this region to be between 13,000 and 33,000 light years from the galactic center and refer to it as the "Galactic Habitable Zone" or "GHZ." Not surprisingly, our Sun lies in an optimum spot in the GHZ, at about 27,000 light years from the center.

Distance, however, is not the only consideration in determining the GHZ. The galactic arms, which stretch from the center to the farthest reaches of the galaxy, are another area to be avoided. The arms are areas of intense radiation and Oort Cloud disturbing gravitation. Moreover, they are dense with gases and interstellar matter, making them the birthplaces of stars and the sites of frequent supernovas. As a result, the arms are as dangerous as the galactic center.

As you might guess, our Sun lies just outside the galactic arms, though this in and of itself is no guarantee of safety. That is because the entire galaxy rotates. Since rotation is not the same in all places, even stars lying outside of the arms must occasionally pass through these danger zones. Luckily, our Sun rotates at nearly the identical speed as the arms, minimizing the number of times they have to be traversed. Also, the Sun's galactic orbit is nearly circular, rather than elliptical as is the case with most stars, further minimizing the number of such crossings.

Astronomers estimate that more than ninety-five percent (perhaps significantly more) of the stars in our galaxy lie outside of the Galactic Habitable Zone. If so, that would reduce the number of stars that are candidates for the formation of habitable planets to a still prodigious ten billion (i.e., 200 billion [stars in the galaxy] x 5% [stars in the GHZ] = 10 billion).

3. In the Stars

Now that we have our 10 billion stars in the GHZ, the next question, again, is whether some of these stars are more likely to host habitable planets than others. Or are they all pretty much the same?

Well, they are not at all the same. First, there are "main sequence" and "non-main sequence" stars. Main sequence stars are those that are in the prime of their lives and are creating their energy by fusing hydrogen into helium. Non-main sequence stars, which comprise about 10% of all stars, are old stars that have used up their available hydrogen and are fusing helium and heavier elements. As we discussed earlier, this causes the stars to hugely expand and become red giants. Red giants will have already devoured any planets in their habitable zones. Moreover, red giants fairly quickly either explode or implode, leaving a white dwarf or a neutron star, which have ceased fusing altogether, or worse, a black hole. So, I think we can safely exclude non-main sequence stars from our search.

Astronomers divide main sequence stars into seven major types, those being O, B, A, F, G, K and M (a handy mnemonic is, "Oh Be A Fine Girl/Guy, Kiss Me" or the more questionable, "Odd Boys And Funny Girls Kiss Me"). Since the letters themselves have no particular meaning, their use seems a bit arbitrary. They are actually the result of an historical accident. In the 1890s, scientists developed a classification scheme for stars by assigning each star type a letter according to how much hydrogen was observed in their spectra, from A (with the most) to B (with the next most), to C and on down the alphabet. There were 22 types in all. The scheme was cumbersome and of limited use. So, in 1901 Annie Jump Cannon simplified the system into a sequence of temperatures, rearranging and combining the previous twenty-two categories into just seven, leaving us with the odd-looking OBAFGKM classification we use today.

Ms. Cannon classified the stars according to temperature, from hottest (O) to coolest (M), but the classification is equally indicative of size and luminosity. Not surprisingly, big stars are hotter and brighter than little stars. The classification, however, works inversely for lifespan and abundance, with O stars the least abundant and shortest lived, and M stars the most abundant and longest lived. A simplified chart of star characteristics is set forth below for reference:

Type	Color	Temperature	Size	Luminosity	Life Span	Relative Abundance
O	Blue	≤25,000	60x	30,000x	2-8 million years	0.000003%
B	Blue	10,000+	10x	25,000x	100 million years	0.13%
A	Blue	7,500+	3.2x	15.0x	400 million years	0.6%
F	White	6,000+	1.5x	3.5x	5 billion years	3.0%
G	Yellow	5,000+	1.1x	1.2x	10 billion years	7.6%
K	Orange	3,500+	0.8x	0.4x	15 billion years	12.1%
M	Red	≥3,500	0.3x	0.04x	100 billion years	76.45%

NOTE: Surface temperatures in Kelvin; Size and Luminosity in comparison to the Sun.

So, which stars are the most hospitable to potentially habitable planets?

I think we can eliminate the big blue stars (types O, B and A) right off the bat. Because of their great mass, the squeeze of gravity is enormous, causing the star's internal pressures and temperatures to soar. This allows for the highly efficient fusion of hydrogen into helium but results in the stars using up their hydrogen reserves quickly, shortening their lives to only a few tens of millions of years. It took our solar system more than half a billion years just to settle down to the point where life could begin to take root. At a maximum of 400 million years, the blue stars' lives are likely too brief for life to begin, much less to evolve technological societies.

The big stars comprise less than one percent of all stars, so no big loss. Let's move on, this time to the other side of the spectrum.

Our experience here on Earth indicates that life evolves slowly. Life has exited here for perhaps as long as four billion years, but advanced technology for only about one hundred years. So, a stable habitat seems to be an essential prerequisite for complex life. M-type or red dwarf stars (not to be confused with red giants) are miserly with their fuel, burning it at a very slow rate. Furthermore, red stars are fully convective, meaning hydrogen is constantly recirculated from the star's outer regions to its core, enabling them to burn their entire supply of hydrogen. Stars like the Sun, on the other hand, are not fully convective and are able to burn only the ten percent of their hydrogen that is located in their cores. Consequently, red stars will live for a hundred billion years (our Sun, by comparison, will last only ten billion years). That's about as stable as it gets. Moreover, red dwarf stars are by far the most abundant, comprising over three quarters of all stars. There are numerous red dwarfs within the neighborhood of our own Sun, including the nearest, Proxima Centauri, which is a mere 4.244 light years away. Indeed, of the 63 stars within five parsecs of our Sun, fifty are red dwarfs. Interestingly, however, none of these stars are visible to the unaided eye because of their small size and low luminosity. Anyway, their longevity and abundance would seem to make the M stars very promising candidates for the evolution of technological life.

There are problems, however. For one, their small size and resulting low levels of energy output mean that a red dwarf's Circumstellar Habitable Zone, i.e., the area around a star where water on planets can remain in a liquid state (which we will get to in a moment) is very near the star. When a planet orbits closely to its star, the star's gravity gradually slows the planet's rotation until it stops completely. Thereafter, only one side of the planet will face the star, and the planet will be said to be "tidally locked" with its star. The same process has occurred with our own Moon (which always shows us the same face) and with every other large moon in our solar system.

When only one side of a planet faces its star, the facing side becomes superheated, while the dark side becomes frigid, perhaps cold enough to freeze its atmosphere. Some scientists have postulated possible mitigating factors, such as heat transfers from the facing to the dark side by winds or ocean currents, but it remains that life will have difficulty evolving intelligent life forms, or emerging at all, on tidally locked planets.

Another problem with M-type stars are solar flares. A solar flare is an intense, explosive burst of radiation that results from a sudden release of a star's magnetic energy. It has been said that the amount of energy released in a flare can be equal to millions of hundred megaton bombs exploding at the same time.

M stars tend to rotate rapidly, creating very strong magnetic fields, which causes frequent flares. These flares can be many times more intense that those from our Sun. Any closely orbiting planet in a red dwarf star's habitable zone would likely be continuously scorched by such flares, sterilizing its surface.

Finally, there is the solar wind. This is the stream of charged particles that are constantly emitted into space in all directions by a star. The solar or stellar wind can rapidly erode a planet's atmosphere, and the radiation can cause severe damage to any life on the planet. Earth's magnetic field protects us from our Sun's solar wind by redirecting it around the planet. The magnetic field is created by the Earth's nickel iron core and the spinning of the planet, a process referred to as the geodynamo mechanism. The problem is that a planet orbiting a red dwarf in its habitable zone, as we have discussed, will likely be tidally locked, meaning that, instead of rotating daily as the Earth does, it rotates annually. Such a slowly rotating planet is not going to produce much of a magnetic field, leaving it at the mercy of its star's solar wind. That is not a good situation for life.

All things considered, the red dwarf or M-type stars are not likely to host planets with complex life.

With the biggest and the smallest stars eliminated, let's move on to the next biggest, the F-type stars. These white to bluish white stars are a bit larger than our Sun, having about one and a half times its mass. From our prospective, the primary problem with these stars is their relatively short life span, up to five billion years, but more typically two to four billion. This might seem a sufficient time for life to evolve. Indeed, our planet is only four and half billion years old, and there has been life here for between three and a half and four billion of those years. Planets, however, do not remain habitable for the entire life of a star. Conditions on the early Earth (the first half to three quarters of a billion years), were not conducive to life. Although our Sun's life expectancy is ten billion years, its characteristics will change as it ages, gradually growing larger and hotter. In fact, life on Earth will likely come to an end within perhaps another half billion years, or at about half of the Sun's life expectancy. That means life on an F star has only a small window of opportunity to evolve intelligent or even multicellular life, probably insufficiently small, if life on Earth is our guide.

Moreover, F stars emit far great quantities of high energy light, such as UV radiation, than stars like our Sun, which can cause profound damage to DNA molecules. It has been experimentally shown that a planet orbiting an F-type star in an orbit equivalent to the Earth's (being a bit further out to adjust for the larger, hotter star) could expect UV radiation to cause between two and a half and seven times as much damage to DNA as it does on Earth. Furthermore, this intense UV light would seriously degrade other hydrocarbon molecules essential for life. To survive on such a planet, life would have to be protected against such radiation by a significant ozone layer or retreat under water or underground, where the development of technology would be unlikely.

So, another classification of stars is eliminated, though F stars comprise only about three percent of all stars. That leaves us with the G-type, such as our Sun, and the K-type or orange dwarfs.

Some scientists think that K stars may be the best places of all to search for extraterrestrial life. Their relatively long lifespan (at least fifteen billion years as a main sequence star, compared to our Sun's ten billion) offers the long-term stability that appears to be necessary for the establishment and development of life. Moreover, there are quite a lot of them, K stars comprising about one-eighth of all main sequence stars.

Because of their smaller size, however, the habitable zones around K stars are very near in. Thus, just as with the M stars, tidal locking is a problem for K stars, particularly for smaller K stars. Since the jury is still out with regard to what percentage of K stars will have tidally locked, habitable zone planets, we will, for purposes of this paper, generously accept all K Stars.

Our experience here on Earth strongly indicates that intelligent life can (and maybe one day will) arise on G stars, such as our Sun. So, we will take all of the G stars as well.

G and K stars comprise (at most) twenty percent of the main sequence stars, which figure we will use for our computations. With a bit of rounding, we have now narrowed our search to about one percent of the stars in the galaxy or two billion stars (i.e., 200 Billion [stars in the galaxy] x 5% [GHZ stars] x 20% [G and K stars] = 2 Billion).

4.   Evil Twin

A great many stars do not travel alone. Such stars have a companion star (sometimes two or three) and are referred to as binary star systems. There is much disagreement as to how many stars are in binary relationships with other stars. That is probably because the number varies depending on star size. Eighty percent of massive stars (having more than two solar masses, i.e., twice as large as our Sun) are binary, while only twenty-five percent of small stars (having less than 0.5 solar masses), such as red dwarfs, are so encumbered. Half of all G and K stars, which are the stars we are interested in, are in binary systems.

The question for us is the effect that binary systems might have on the planets orbiting such stars. That could depend on the proximity of the two (or more) stars. If the stars are very close, closer to each other than the orbit of mercury around our Sun, the planets might orbit both stars as if they were a single star. It is not clear, though, whether such an orbit would be stable, and, in any event, such very close orbiting stars are probably limited to red dwarfs, which are not our concern here.

Many binary stars orbit each other within the radius of our solar system. In those cases, a planet orbiting one of the stars will be tugged back and forth by the gravity of the other star, making the planet's orbit highly irregular. Such a planet would be pulled in and out of a habitable zone, probably making life impossible. Moreover, many such planets would be cast out of the solar system and into the frozen wastes interstellar space. Not a good place for life.

A significant number of binary systems involve stars that are very far apart, as much as a thousand times the distance separating the Earth from the Sun. One might think that those systems are more stable since the companion star is so far away that its gravity on the other's planets would be negligible. However, a new study shows that may not be the case. Astrophysicists, Nathan A. Kaib, Sean N. Raymond and Martin Duncan, ran extensive computer simulations to model exoplanets residing in wide binary systems. They found that perturbations from other stars outside the binary system had a profound effect on the shape of the system's orbits, with the effect that planets were either ejected from the system entirely or ended up in highly eccentric, elongated orbits.

It appears then, that binary systems, no matter how close or distant the companion stars, will interfere with planet formation and likely not permit the stable planetary orbits needed for development of life. But don't take my word for it. As astrophysicist, Paul Sutter, recently put it:

"While binary systems certainly have a habitable zone, where liquid water could potentially exist on the surface of a planet, life might find it difficult to gain a foothold. Orbiting two stars at once, as our friend Kepler-47c does, makes life very elliptical, occasionally bringing the planet out of the zone. Life doesn't take too kindly to frequently freezing over. Orbiting just one star in a binary system? Well, sometimes you'll have two stars in your sky at once, which can be a tad toasty. And sometimes you'll have a star on each face of the planet, ruining the night. And don't forget the double-doses of UV radiation and solar flares. With that kind of instability, erraticism and irradiation, it's hard to imagine complex life evolving with the kind of regularity it needs."

Not particularly well put, perhaps, but the point is clear; life would have a difficult go of it in a binary system. So, we will eliminate binary stars from our calculations, reducing our candidates to one-half of one percent of the stars in the galaxy, though a still sizable billion stars (i.e., 200 Billion [stars in the galaxy] x 5% [GHZ stars] x 20% [G and K stars] x 50% [non-binary stars] = 1 Billion).

With our billion suitable stars selected, let's begin our search for suitable planets.

5.   Oddball

The Philosophers. In his 4th Century B.C cosmological work, *On the Heavens*, Aristotle proposed a geocentric model of the universe, where a spherical Earth was surrounded by fifty-five concentric, crystalline spheres to which were attached the Sun, the Moon, the planets and the stars, all of which rotated in perfect, circular paths around the Earth.

The second century Alexandrian astronomer, Claudius Ptolemy, described this geocentric model mathematically in his masterful (and largely incomprehensible) astronomical treatise, the *Almagest*. Ptolemy's model remained the accepted model of the universe for about fifteen centuries, when Nicholas Copernicus devised the heliocentric model, replacing the Earth with the Sun at the center of the universe, while maintaining Aristotle and Ptolemy's perfectly circular planetary orbits.

The problem was that both the Greek and the Copernican models almost but did not quite accord with actual observations of the celestial bodies. In particular, they could not explain the apparent retrograde motion of the planets (where a planet appears to start moving backward in its orbit because of the relative positions of the planet and the Earth as they move around the Sun). As a result, the models had to be continuously adjusted through ad-hoc offsets, such as epicycles (a smaller circle turning on a larger circle, like petals on a flower), to make the models correspond with reality. The system worked well enough for the day, but it was difficult and cumbersome.

As a result of his having briefly served as the assistant to the meticulous Sixteenth Century Danish observational astronomer, Tycho Brahe, the mathematician/astronomer/ astrologer, Johannes Kepler, gained access to decades of Brahe's precise observational data on the motion of the planets. Kepler's long and careful study of the data led him to conclude that the orbits of the planets were not circular, as had been supposed for the previous couple of thousand years, but, instead, were elliptical, with the Sun at one foci (an elliptical orbit having two foci, being the two points around which the object orbits). In his three laws of planetary motion, Kepler was able to describe the planetary motion observed and recorded by Tycho Brahe.

Although Kepler's laws provided a good approximation of the motion of the planets, they were not completely accurate, primarily because they failed to take into account the effects of gravity (of which Kepler knew nothing). That's where Newton enters the story. Sir Isaac gave us his three laws of mechanics, those being inertia (objects in motion stay motion), reaction (every action has an equal, opposite reaction) and motion (the acceleration of an object is proportional to the force acting on that object). He also, by the way, gave us his Universal Law of Gravitation (the gravitational force between two objects is proportional to their masses and diminishes by the square of the distance between them). These laws explained mathematically the movement of the planets in our solar system as described by Kepler, but to a far greater degree of certainty.

Finally, in 1905 Albert Einstein presented his theory of special relativity. The theory explains differences in motion, mass, distance and time that result when objects are observed from two radically different frames of reference, which Newtonian laws could not. Since few things move at speeds fast enough for us to notice relativity, Newton's laws can be and are still used to describe planetary motion. For this reason, many scientists see Einstein's theory of relativity not as a replacement of Newton's laws of motion and universal gravitation, but as their culmination.

As a result of the work of these and other thinkers, we now know how the planets move and the solar system works. But how did it get here in the first place?

Planetary Formation. Stars form in vast clouds of gas (hydrogen and helium) and dust (heavier elements created within earlier stars) termed nebula (Latin for cloud). Because nebula may spawn numerous stars, they are often referred to as stellar nurseries.

Turbulence within a nebula can cause gas and matter to aggregate. When such an aggregation reaches a sufficient mass, it begins to collapse under its own gravitational attraction. As more and more dust and gas collects, it begins to rotate. Most of the matter (perhaps 99%) forms into a sphere at the center, termed a protostar, and the rest flattens out through the action of gravity into a disk, termed a protoplanetary disc.

4.6 billion years ago, such a collapse occurred within a nebula in one of the galactic arms of our galaxy, out of which our solar system would form. Since the universe came into existence 13.8 billion years ago and our galaxy about 13.6 billion, both had already been going concerns for some 9 billion years by that time. As gravity contracted the central sphere, the internal pressure and temperatures rose, eventually reaching the point where nuclear fusion began.

At the same time, bits of dust in the protoplanetary disk circling the central sphere began to coalesce through random collisions and electromagnetic attraction. Over time, these aggregations of matter grew to the size of grains of sand, then pebbles, then boulders. When they reached about a kilometer in diameter, they became large enough to attract matter through gravity, rather than simply by chance collisions. Such larger objects are called planetesimals. In their orbits around the new Sun, the planetesimals continuously collided and coalesced with debris and even other planetesimals, creating ever larger objects. When they reached about the size of the Moon, they became what are termed protoplanets. Radiation and the energy released from so many collisions heated and melted the matter that formed these objects. The gravity of the protoplanets pulled the molten materials inward toward their centers, forming spheres. Some of these spheres eventually cleared their orbits of all other protoplanets, planetesimals and debris, leaving only eight, those being the planets of our solar system.

There was a big difference between the formation of planets close to the Sun and those further out. At a distance somewhere between the present orbits of Mars and Jupiter, it was cold enough for volatile compounds, such as water, ammonia, methane and carbon dioxide, to condense into ice grains. Astronomers refer to this point as the snow line of the solar system. Out beyond the snow line, the newly forming planets vacuumed up the ice in addition to the bits of matter, allowing them to grow ever larger. When the mass of such a protoplanet reached about ten times that of the present Earth, it would have had enough mass to gravitationally attract and begin absorbing the vast amounts of hydrogen and helium in the disc, causing them to expand into the giants of today.

In recent times, scientists have come to realize that the outer giant planets in our solar system come in two varieties. Jupiter and Saturn are truly huge in comparison to the Earth, Jupiter having about 320 times the mass of the Earth and Saturn 95. They are composed mostly of hydrogen and helium (90% by mass) and are referred to as gas giants. Uranus and Neptune, on the other hand, are a good deal smaller, Uranus being about 15 times Earth's mass and Neptune 17. They are largely rock (25% by mass) and ice (frozen water, methane and ammonia) (60%), with only a relatively small amount of hydrogen and helium (15%). They are now known as ice giants. The difference in composition between the two types of giant planets is likely the result of there simply having been much less hydrogen and helium in the outer reaches of the protoplanetary disc for Uranus and Neptune to absorb during the early days of planet formation. With only limited quantities of these gases available, Uranus and Neptune could not grow to the size of the gas giants.

Closer in, ice was not available, and thus the planets did not grow so large. Moreover, when fusion kicked in and the Sun switched on, it began to radiate plasma and charged particles (electrons and protons). This solar wind swept away the hydrogen and helium from the regions closest to the Sun, leaving only heavy, rocky materials. Further away, however, the winds were not so strong, allowing these gases to coalesce around the newly forming planets.

And so, we have the relatively small, rocky inner planets, Mercury, Venus, Earth and Mars, and the outer giants, Jupiter, Saturn, Uranus and Neptune. Astronomers refer to this process of planet formation as the planetary accretion model.

By about four billion years ago, having devoured essentially all of the debris in their paths, the planets settled into their orbits, where they have remained ever since. Though not perfect circles or eternal, as envisioned by the early astronomers, the planetary orbits are regular and unchanging, allowing for the long-term stability of the planets that would seem to be required for the development of life.

Long before the discovery of exoplanets (i.e., planets orbiting stars other than our Sun), astronomers strongly suspected their existence. After all, our star formed within a nebula just as did all the other stars, and it seemed a logical certainty that the accretion model of planetary formation would be equally applicable to all such other stars. Thus, when the search for exoplanets began, it was expected that any planetary systems found would look a lot like our own, with small rocky planets close to the system's star and gas and ice giants farther out, all orbiting in nearly circular, stable orbits, in perfect harmony. That, however, would not be the case.

Exoplanets. Exoplanets are hard to detect, primarily because they are small and very, very far away. How far away? Well, scientists often use analogies to help lay people conceptualize great age, size or distance. I'll try my hand at that.

Imagine, if you will, that the Sun is the size of a pumpkin. In that event, the Earth would be the size of the head of a pin. Do you know how far the pinhead would be from the pumpkin? Three feet? Five feet? Try 50 feet. Now, try to guess how far the next closest pumpkin to our pumpkin would be? On second thought, don't even bother. It would be more than 5,000 miles away. That's about the distance from St. Louis to Moscow. As they say in pumpkin country, that's a fer piece. I reckon it would take a pretty powerful pair of field glasses for a who on a pinhead to spot another pinhead circling a pumpkin 5,000 miles away. To complicate matters, imagine that the pumpkin is glowing with the intensity of the kind of flashbulb that leaves a blue spot in your vison for a couple of weeks after you say cheese. Just try to pick the pinhead out from the glare.

So, you can see the enormity of the task astronomers set for themselves when they set out to find exoplanets.

The Search Begins. Because stars are so far away, they appear, as they do from your backyard, as dimensionless points of light, even with the most powerful of telescopes. Any planet orbiting a star would be merged into this point of light, making it impossible with today's technology to directly view an exoplanet. And, even if we could, the dim image of the planet would be lost in the glare of the star. Although we cannot directly view exoplanets, astronomers have devised ingenious methods of indirectly detecting these planets by inference from observations of their host stars.

One important method is known as radial velocity or Doppler spectroscopy. This method relies on the fact that a star when orbited by a planet will move in a little circle or ellipse in response to the gravitational tug of the orbiting planet. When the planet is nearest to the observer, the star is pulled ever so slightly in the observer's direction. Conversely, when the planet is on the other side of the star, it pulls the star a tiny bit away. Light, like sound, can be thought of as a series of waves. When an object approaches an observer, the waves reach the observer more quickly than they would if the object were stationary. This causes the waves to start piling up, shortening the distance between one wave and the next. This compression of the sound waves increases their frequency, causing the pitch to rise, as with the sound of an approaching car. When the object recedes, however, the length between the waves stretches out, lowering the pitch, as when the car passes by. This is known as the Doppler effect, and it affects light in the same way as it does sound. When an object approaches, the light waves are compressed, shortening the distance between the waves and making the light bluer (blue being the shortest wavelength of visible light). The light waves from a receding object become longer, shifting the light toward red (the longest wavelength). Using highly sensitive spectrographs, scientists can track a star's light spectrum, watching for periodic shifts toward the red then the blue, almost certainly indicating the presence of an orbiting object. This method allows scientists to learn not only the existence of the orbiting planet but much about its mass, location and movement.

The other primary method of detecting exoplanets (as of this writing) is transit photometry. When an orbiting object crosses between an observer and a star, the object blocks some of the starlight. Thus, when the moon crosses in front of the Earth, the Sun's light can be all but obliterated. Similarly, when Venus passes between the Earth and the Sun, a little black silhouette of the planet can be observed traversing or transiting the Sun. When an exoplanet transits its star, the black dot of the planet cannot actually be seen. The starlight, however, will be dimmed slightly, by the amount of light blocked by the planet. This enables scientists to infer the mass of the transiting planet.

Using both of these methods in conjunction can provide a great deal of information as to a planet's size, mass, orbit, distance from its star and even its composition. The development of these methodologies enabled astronomers to begin the search for exoplanets in earnest.

In 1995, the Swiss team of Michel Mayor and Didier Queloz made the first confirmed discovery of an exoplanet orbiting a main sequence star. The planet orbited the star 51 Pegasi and, consequently, was named 51 Pegasi b. The planet was declared, by those who declare such things, an oddball. It was odd because there was nothing like it in our solar system and it deviated from our understanding of planet formation in two respects. First, the planet orbits very close to its star (less than five million miles out) and completes its orbit in just 4 days. By comparison the Earth's average orbital distance is about 94 million miles and its orbital period is 365 days; Venus' are 67 million miles and seven months; and Mercury's 37 million and 3 months. As a result of its close proximity to its star, the surface temperature of 51 Pegasi b is about is 1000 degrees K (1800 F). What is truly odd is that 51 Pegasi b is a gas giant that is more than one and a half times as large as Jupiter. The discovery baffled astronomers. A gas giant planet nestled right next to its star was completely at odds with the accretion theory of planetary formation, which posits that a gas giant can form only far away from its star, out beyond the snow line. But there it was. The very first exoplanet apparently blowing a hole through the long-accepted understanding of planetary formation.

The Swiss made their discovery using the radial velocity method, which looked for wobbles in a star's spectrum, indicating the presence of an orbiting object. Astronomers had not expected large planets orbiting closely around their stars. So, once the finding had been confirmed, other astronomers began re-examining their own radial velocity observations. Low and behold, giant star-hugging planets began popping out of the data. Further observations revealed more of these oddballs, first by the dozens, then by the hundreds.

Hot Jupiters, as these close orbiting giants began to be called, dominated the early years of exoplanet exploration, largely because the radial velocity method was the only practical means of discovery at the time and large, close orbiting planets cause more detectible wobble than smaller, more distant ones. This changed with the launch of NASA's Kepler Space Telescope in 2009. Kepler used the transit method, watching for a star to dim as a planet passes in front of, or transits, the star. From its position in space, Kepler was able to detect smaller planets.

Kepler was a highly successful venture, discovering more than 2,700 confirmed exoplanets. Interestingly, the most common type of planet discovered by the mission was another oddball, the so-called super-Earths or mini-Neptunes. These are planets having masses between that of Earth and that of Neptune, a class of planets that we do not have in our solar system.

In our solar system, there is a range of planetary sizes and distances between planets. In most of the other planetary systems we have encountered, however, the planets have very similar sizes and regular spacing between their orbits, like peas in a pod, as some astronomers put it. Moreover, these pea pod planets orbit very close together, unlike the planets in our solar system that have widely spaced orbits, and most of them orbit very near their star. By way of example, the three super-Earth systems orbiting the star GJ 9827 located 100 light years from Earth have radii of 1.6, 1.2 and 2.1 times the radius of the Earth, and all orbit the star within 6.2 days. Another, the Trappist-1 system, consists of seven similar sized planets orbiting very closely together, two of which orbit their star at a distance of only about one and half times the distance from the Earth to the Moon. The first planet of the system orbits its star in 1.5 days while the seventh in only 18.8 days (remember, Mercury takes 90 days to orbit the Sun). I could go on.

And then there is a class of planets having highly elliptical (eccentric) orbits. One, named HD 80606 b, is about the size of Jupiter, though four times as massive. It spends most of its 100-day year travelling an oblong route away from then back towards its star. Then, in a matter of about twenty hours, it sweeps around the star, very nearly touching it, before heading back out again. The most eccentric of all is HD 2078 b, a Jupiter sized planet with a measured orbital eccentricity of 0.96, meaning that its orbit is a nearly flattened ellipse, very much like the orbit of a comet (Haley's Comet's eccentricity being 0.97). At the farthest from its star, it is about two and half times the distance of the Earth from the Sun. At its closest, only 0.06 the distance. Very strange. You can imagine what would happen to the other planets in our solar system if Jupiter had such an orbit.

Planetary Migration. Hot Jupiters, mini-Neptunes, super-Earths, peas in a pod, planets orbiting right next to their stars, planets in highly eccentric orbits. None of this was envisioned by the accretion model of planetary formation. So why was the theory so wrong? Well, maybe it wasn't. Maybe it just didn't go far enough. The model is still the generally accepted description of planetary formation, but it doesn't address what happens after the planetary system is formed.

When the inertia (forward movement) of an object in space moving away from another object is stronger than the gravitational attraction of the two objects, it will speed on by. When the inertia is weaker, the object will be pulled into the other. When the inertia and the gravitational attraction are equal, the object will orbit the other. Newton's first law of mechanics tells us that an object in motion will stay in motion unless acted upon by some other force. The planetary accretion theory assumes that a newly forming planet will clear its orbit of dust and gas. Since there would be nothing left in the orbit, the planet should continue to orbit the star forever (or at least for the billions of years until the star leaves the main sequence). Recent research, however, shows that this may not be the case.

As you will recall, planets form from the gas and dust in the protoplanetary disk orbiting a newly formed star. As a forming planet plows through this material, drag causes the planet to spiral inward toward the star, just as Earth's atmosphere causes orbital decay of satellites in low Earth orbit. As the planet nears its star, tidal or gravitational forces acting on the planet cause the planet's orbit to become more circular and thus more stable, bringing the inward movement to a halt. Planets can remain in such close-in orbit for billions of years. This planetary migration appears to be intrinsic to most planetary systems.

The Grand Tack. If planet migration is such a common phenomenon, I'm sure you are asking yourself, why didn't it happen in our solar system? Well, actually, it appears that it did.

To explain why the solar system appears to be so different from other planetary systems so far encountered, with the absence of super-Earths and mini-Neptunes, gas giants orbiting far away from the sun, no planets within the orbit of Mercury and all planets in stable, spacious, nearly circular orbits, astronomers have devised several models of the early solar system, most famously the Grand Tack and Nice Models. In these models, Jupiter began migrating toward the Sun very early (the first five million years) in the process of planetary formation. At the time, the inner solar system was populated by rocky planets in the process of becoming super-Earths. As Jupiter moved inward, its gravitational pull would have thrown these planets into close, overlapping orbits, setting off a series of collisions. The avalanche of debris from the collisions raised powerful aerodynamic interference in the disk, forming spiraling swirls of gas that swept the inner planets into the Sun. At the same time, our other gas giant, Saturn, also began migrating toward the Sun. Eventually, Jupiter and Saturn became temporarily locked in a 3:2 orbital resonance, meaning that for every three orbits of the Sun by Jupiter, Saturn made two. The push and pull of the two planets gradually moved them away from the Sun, and eventually they settled into their present orbits. The commotion caused by these wandering giants, by the way, wreaked further havoc in the solar system, such as propelling Neptune past Uranus and preventing the asteroid belt from accreting into a planet. With the inner planets gone, three new small rocky planets, Mercury, Venus and Earth, formed from the leftover materials pushed into the inner solar system by Jupiter's approach. This created a second generation of inner planets (Mars, apparently, being the only one of several protoplanets in its orbit that managed not to be swept into the inner solar system by Jupiter's intrusion).

Oddball. So, there you have it. After finding that the oddball hot Jupiters, super-Earths, mini-Neptunes, closely orbiting planet groups and star hugging planets are common, indeed the norm, we find that our solar system is actually the oddball. Not at all expected.

Is our system unique? Rare? Or just uncommon? It's hard to say. We haven't yet found another one quite like ours, though that may be only a matter of not having the technology to do so at this time. If the models showing that the inner planets are actually second-generation planets resulting from a serendipitous cascade of events are correct, I think it probable that our solar system is a rare thing, on the order of one in a thousand or fewer. But, even keeping our estimates conservative, I think it unlikely that our solar system is representative of more than one percent of all systems.

Doing the math, this leaves us with ten million proper stars with stable, well-positioned, well-spaced planetary systems amenable to technological life (i.e., 200 Billion [stars in the galaxy] x 5% [GHZ stars] x 20% [G and K stars] x 50% [non-binary stars] x 1.0% [stable planetary systems] = 10 Million).

6.   Habitability

In this section we will search for habitable planets. By habitable, I mean planets capable of sustaining not just life, but intelligent life with the potential for developing technological civilizations. The following are at least some of the considerations.

The Liquid Water Habitable Zone. Life is chemistry. For the most part, chemical reactions occur in liquids, where, unlike solids, molecules have freedom of movement and where, unlike gases, molecules can be concentrated. Liquid, however, is the least common state of matter in the universe. This is largely due to the limited range of temperatures and pressures that allow substances to remain in a liquid state.

The three most abundant liquids (other than those found deep within a planet, such as liquid hydrogen and molten lava, neither of which are particularly conducive to life) are water, methane and ammonia. We know that water can sustain life, but what about methane or ammonia? On Saturn's moon, Titan, it rains liquid methane which collects in vast lakes. Could life arise in a methane lake? Perhaps. But there is a problem. Both methane and ammonia remain liquid only at very low temperatures, methane, at one atmosphere of pressure, from -164 C to -182 C (-263 F to -295 F) and ammonia from -33 C to -78 C (-28 F to -108 F). Such cold temperatures slow the rate of chemical reactions to a snail's pace, which is not a pace amenable to living creatures, other than, perhaps, snails.

Water (H_2O, that is) is common throughout the universe, being composed of the most abundant element, hydrogen, and the third most, oxygen. Moreover, it remains a liquid at relatively high temperatures and is an excellent ("universal") solvent for carbon-based molecules. So, I think it reasonable to conclude that a habitable planet must have abundant liquid water.

For there to be water, the planet must be warm, but not too warm, which requires it to be close, but not too close, to its star. The region of a solar system where liquid water can exist is termed the circumstellar habitable zone (not to be confused with the galactic habitable zone discussed previously). Since there are other types of habitable zones, we will sometimes refer to it as the liquid water habitable zone.

The location of a planetary system's liquid water habitable zone depends on a number of factors, such as the luminosity of the star, the size of the planet and whether and what type of atmosphere and clouds the planet has. Thus, the habitable zone will vary from planet to planet.

It would seem reasonable to suppose that a star with multiple planets might very well have at least one planet within the habitable zone. However, it's a bit more complicated than that. The location of a star's habitable zone changes over time. That is because a star will become more luminous as it ages. Consequently, a planet might start out in the habitable zone but later, as its star gets hotter and the habitable zone move outward, it might find itself left out in the cold, or rather, in the heat.

That is precisely what happened to Venus. Four billion years ago, the Sun was thirty percent less luminous than it is today. At that time, Venus was likely just within the inner edge of the habitable zone and might well have been a C.S. Lewis Perelanda-like tropical paradise. As the Sun got hotter, however, its oceans evaporated, pumping vast quantities of water vapor (an excellent greenhouse gas) into the atmosphere, trapping the Sun's heat and causing temperatures to skyrocket. Over time, the water vapor molecules were broken apart by ultraviolet radiation, allowing the hydrogen to escape into space. With no water on the surface, carbon dioxide built up in the atmosphere, leading to runaway greenhouse conditions. Today, temperatures on the surface of Venus approach 900 degrees Fahrenheit, the atmospheric pressure is ninety times that of the Earth and the atmosphere is filled with toxic, corrosive gasses. It is a literal hell on Earth, or rather, on Venus. A sad story.

Earth, on the other hand, was much luckier ... quite fortunate, in fact. You see, if the Sun were as dim today as it was four billion years ago, the Earth would be a frozen snowball. Our present, more robust Sun now provides plenty of heat to keep the oceans liquid and life chugging along. Yet, the evidence is that Earth was warm and had liquid oceans four billion years ago, even though the Sun was much dimmer. How could that be? This is referred to as the Faint Young Sun Paradox. Much controversy surrounds the subject, but the most plausible explanation is that there were significantly more greenhouse gases in the atmosphere at that time (perhaps several hundred or a thousand times more carbon dioxide, vast amounts of water vapor and high levels of methane as well). Over the eons, the greenhouse gases diminished while the Sun's luminosity increased, keeping the Earth in the sweet spot of the habitable zone. Remarkable.

Of course, the Sun keeps getting hotter and the habitable zone keeps moving further out. It is estimated that in as little as half a billion years it will leave us behind, and Earth will join Venus in hell.

At about that time, Mars will enter the habitable zone. If it were the type of planet where life could flourish (which it is not), it might cast off its icy shroud and bloom for a short interval, at least until the Sun leaves the main sequence and roasts the inner planets.

A movable habitable zone is a significant problem for technological life. Whether a planet starts inside the zone and is left behind or starts outside and then enjoys a period within, the movement significantly limits the time it will be habitable. Because of its serendipitous placement, Earth had managed to remain within the habitable zone for four and a half billion years, and it has needed every one of them to produce a technological civilization. Like much else about our planet, that would seem not to be the norm.

The UV Habitable Zone. Solar radiation is radiant energy emitted by the Sun. From longest to shortest wavelength, it includes infrared, visible and ultraviolet radiation. Most of the solar radiation that reaches Earth is made up of visible light and infrared light. Only a small amount of ultraviolet light reaches the surface.

UV radiation, however, is generally considered dangerous stuff. It can inhibit photosynthesis, induce DNA destruction and damage a wide variety of proteins and lipids. On the other hand, UV is necessary for the synthesis of many biological compounds and macromolecules necessary for life and may have been crucial to the formation of RNA on the primitive Earth.

As with liquid water, there is a zone around a star where a planet will receive enough UV radiation for the synthesis of biochemical compounds but not so much as to damage biological systems. As you might have guessed, this zone is referred to as the UV habitable zone.

In a recent study, researchers calculated the inner and outer UV habitable zones of stars having masses ranging from 0.08 – 4.0 solar masses (i.e., the mass of the Sun), then overlaid the UV habitable zone on to the liquid water habitable zone, the overlap being the place most likely to sustain life. Their results were surprising. They found that only stars having 1.0 -1.5 solar masses have any overlap at all. If correct, that would eliminate most of the orange K stars, which have less than 1.0 solar masses and which comprise more than half of our candidate stars.

Magnetic Fields. The Sun emits a continuous flow of charged particles, primarily protons and electrons, referred to as the solar wind. The particles travel at a million miles per hour and can strip away a planet's atmosphere and oceans.

When the Earth was first formed and very hot, heavy elements sank through the molten rock toward the center, forming a core twice the size of the moon composed mostly of iron. Although the inner core (760 miles thick) is as hot as the surface of the sun (5,200 C/9,300 F), the crushing pressure of gravity keeps it a solid. Surrounding the inner core is an outer core or iron and nickel some 1,355 miles thick. Because pressures there are lower, the outer core remains liquid. Differences in temperature, pressure and composition within the outer core cause convection currents in the molten iron/nickel, as cooler denser matter tends to sink, while warmer, lighter materials rise. Also, the Coriolis effect caused by the Earth's rotation, creates spinning whirlpools. The flow of the liquid metals generates electric currents, which produce magnetic fields. Their combined effect produces one vast magnetic field surrounding the entire planet. The Earth's magnetic field protects us from the solar wind by channeling the charged particles around the planet.

Since atmospheres and oceans are pretty important for living creatures and since you can't have atmospheres and oceans without a magnetic field, a planet without one is unlikely to be habitable. It would seem that planets with magnetic fields would be fairly common, given that planet formation involves the accretion of rocky materials, with iron sinking to the center to form a core and the whole thing heating up by the huge pressures and naturally occurring radioactive decay. Yet, in our own solar system, the Earth is the only inner planet with a magnetic field of any significance.

Mars had a magnetic field when it was young and its iron core was molten and convecting. It likely also had a reasonably thick atmosphere as well as oceans. Mars, however, is smaller than the Earth, and early on its molten core froze up (or mostly so), turning off the magnetic field and leaving its atmosphere and oceans to the mercy of the solar wind.

Though the same size as the Earth, Venus has no magnetic field. This is apparently due to its glacially slow 243-day spin, which is probably insufficient to rotate the liquid, metallic portion of its core fast enough to generate a magnetic field. Recent studies, though, point out another possibility. Both the Earth and Venus were formed by the collision of countless planetesimals. Iron delivered by these collisions sunk to the middle to form the core. Iron arriving from later impacts, however, had to sink through the protoplanet's mantle, picking up lighter elements, such as oxygen, silicon and sulfur. As a result, the cores developed layers or shells which prevented the generalized circulation of the liquid metals necessary for magnetic fields. The Earth, however, collided with a Mars-sized object late in this process, from which the Moon was formed. This colossal impact may have broken the shells, permitting a generalized circulation. Another lucky break for the Earth.

Although it has no magnetic field, Venus has managed to retain its atmosphere. This is probably because its atmosphere is extremely dense. In fact, the atmospheric pressure on the surface of Venus is equivalent to the pressure at a depth of 3,000 feet beneath the Earth's oceans. That kind of atmosphere will take the solar wind a good while to erode away.

Size. Does size matter? Yes, especially when you're talking about planet habitability. Why? Well, it's all about the atmosphere. The Earth's atmosphere is thin and wispy, making up only 1/1,200,000th of the Earth's mass. Yet, it protects life on Earth by absorbing ultraviolet rays, keeping the surface warm through the greenhouse effect and reducing temperature extremes. A livable planet needs the right kind of atmosphere.

Gravity is what keeps the atmosphere in place (though some of it is continuously escaping into space, which we will get to shortly), and gravity is a consequence of a planet's mass. To hold on to an atmosphere a planet must be big, but as we will see, not too big, and not too little either. Let's start with the not too big.

As we have seen, the vast majority of exoplanets found thus far are mini-Neptunes and super-Earths, which range in size from 2 to 10 Earth masses. When they were first found, it was thought these mid-sized planets were good candidates for habitability. Later, the Kepler mission found hundreds more of these planets using the transit method. This method enabled scientists to determine a planet's radius and orbital parameters but not its mass. Without knowing the planet's mass, it was impossible to determine whether the planet was rocky like the Earth or gassy like Neptune. In recent follow up observations using the radial velocity method (which can measure a planet's mass), astronomers have now obtained the masses for hundreds of these worlds. What they found is the transition from a rocky world to a gaseous world occurs at just twice the mass of the Earth. A planet with two Earth masses in an Earth-like orbit will have enough gravity to hold onto a substantial hydrogen/helium gas envelope, creating atmospheric pressures at its surface hundreds or even thousands of times greater than the pressure at the surface of the Earth. Consequently, mini-Neptunes and super-Earths are much more like Neptune than Earth and are almost certainly uninhabitable.

Small planets, on the other hand, have the opposite problem. Because of their low mass and weak gravity, their atmospheres waste away over time, often very quickly. Mars, for instance, has a tenth of the mass of the Earth and only about 38 percent of its gravity. Once upon a time, water flowed in rivers and lakes under a thick blanket of atmosphere. But its light gravity, along with the loss of its magnetic field, left its atmosphere vulnerable to the solar wind. Today, it is very thin, about a hundred times thinner than the Earth's. Without its insulating atmosphere, Mars is a dry, frozen wasteland. New data from NASA's Mars Atmosphere and Volatile EvolutioN (MAVEN) mission has confirmed that Mars' atmosphere was blown away by the solar wind and is now lost in space. There are many factors that must be considered in calculating the minimum mass a planet must have to retain its atmosphere, but the best estimate I have run across is about one-half that of the Earth.

It appears then, that the mass of habitable planets will range from one-half to twice that of the Earth. Once again, Earth is right in the middle.

Plate Tectonics. With all the talk of global warming and CO_2 levels, do you know how much of the air is composed of carbon dioxide? Go ahead, guess. Wrong. It's .04% (or a little less). In parts per 10,000 (dry air), there are 7,900 molecules of nitrogen, 2,100 of oxygen, 93 of argon and only four molecules of carbon dioxide. That's all. It is a trace component. Once, long ago, CO_2 levels were hundreds of times higher than they are today, and carbon dioxide was a primary greenhouse gas. Not so much any longer. Even at today's miniscule levels, though, it remains of vital importance. That's because carbon dioxide is an essential component of oxygenic photosynthesis (there are other kinds), which produces the oxygen necessary for multicellular life.

The problem is that carbon dioxide is constantly being stripped from the atmosphere. Atmospheric carbon dioxide combines with water to form a weak acid (carbonic acid), which falls to the surface in rain. The acid dissolves rocks and releases various minerals, including calcium, magnesium, potassium and sodium. Rivers carry the minerals to the ocean. There, microscopic calcifying organisms, such as corals, forams and cocolithophores, combine calcium and carbonate (dissolved carbon dioxide) to form, what else, calcium carbonate, which they use to build their shells. When the organisms die, they sink to the seafloor, where the shells accumulate as a sediment that can lithify into limestone. At the same time, phytoplankton in the oceans take in carbon dioxide in the process of photosynthesis. When the plankton and other creatures die, they sink to the bottom into the mud. Over millions of years the heat and pressure compress the mud into sedimentary rocks, such as shale. As time goes on, more and more carbon is removed from the atmosphere and ends up in rocks on the ocean floor. If the level of carbon dioxide drops too low, photosynthesis comes to an end, and with it, complex life. Thus, for there to be complex life, there must be a means of returning the carbon locked away in the rocks on the ocean floor to the atmosphere. That's where plate tectonics come in.

The interior of the Earth can be divided into four layers, a solid inner core, a liquid outer core, a malleable mantle and the outer crust. The crust is cracked into about a dozen major pieces, known as plates, that float atop the mantle. Convection caused by the rising of hot material near the core and the sinking of colder mantle rock causes the plates to move, which is referred to as continental drift. When plates collide, one plate slides below the other back into the mantle in a process called subduction. Oceanic crust is denser than continental crust and will subduct under the continental crust. Under the extremes of heat and pressure, the rock melts, recombining into silicate minerals and carbon dioxide. Volcanos forming along the line of subduction erupt, venting the carbon dioxide and depositing new silicate rock, beginning the cycle again. Because this process can take millions of years to complete, it is called the "long-term carbon cycle."

In our solar system, Earth is the only terrestrial planet where plate tectonics is observed. The others (Mercury, Mars and Venus) have (or had) rigid, spherical shells encompassing the entire planet, with the hot mantle convecting below. This mode of mantle convection is known as "stagnant lid" convection. Stagnant lid planets have no crust subduction and only a short period of volcanic activity, thus no means of re-cycling carbon dioxide, a significant impediment to the development of life.

A recent study has attempted to determine whether plate tectonics is occurring on exoplanets. To do so, scientists needed to know the chemical composition of the planets. With no means of peering into an exoplanet's interior, the researchers looked to the composition of some 1500 stars hosting exoplanets. Since stars and their planets are formed from the same dust and gas, the chemical composition of the stars can be used to model how rocks of varying compositions will react to the high interior temperatures and pressure of a planet. In particular, they wanted to know whether the crust would be sufficiently dense to sustain plate tectonics. They found, to their disappointment, that more than two-thirds of the simulated planets built a crust that was too buoyant for subduction. Obviously, this study is not conclusive, but when coupled with what we know about our own solar system, it appears likely that plate tectonics is uncommon.

For billions of years, plate tectonics and the long-term carbon cycle have kept carbon dioxide levels in the atmosphere consistent and ideal for the development of life on this planet. It is not a perfect system, though. The same process of carbon lithification and burial occurs on the continents as well as on the ocean floor. Continental crust, however, being less dense than oceanic crust, does not subduct. When continental plates ram into each other, mountains are thrust up at the impact site, but neither plate is forced down into the mantle for melting.

Consequently, continental carbon is not recycled, which over time inexorably reduces atmospheric carbon dioxide. The problem is exacerbated by the fact that continents, since they do not subduct, grow ever larger over time. In a billion years or so, it is expected that carbon dioxide levels will be so low that photosynthesis will not be possible. With no plants or oxygen, the lovely greens and blues of our world will wilt into a rusty reddish brown like the surface of Mars. C'est la Vie.

A Big Moon. The Moon is huge, not so much in comparison to all of the moons of the solar system (coming in fifth behind Jupiter's Ganymede, Calisto and Io and Saturn's Titan), but in relation to the Earth. Mars has a couple of city sized rocks for moons, while Venus and Mercury have none at all. The giants have a few sizable moons, but they are tiny in relation to their host. The Earth and the Moon, on the other hand, are nearly a binary planet system. Why is that?

4.5 billion years ago, the planets were still in the process of forming. As the Earth neared its present size, it devoured everything in its path, including objects hundreds of miles across. These encounters would have caused massive explosions but would have had little effect on the much more massive protoplanet. There was, however, another protoplanet in the same orbit as the Earth, which scientists have named Theia (after the Titan goddess who gave birth to the Moon). This was no chunk of space debris but an emerging planet as large or larger than Mars. The inevitable inevitably happened, and Theia smashed into the Earth. The collision annihilated Theia, which vaporized into an immense, incandescent cloud tens of thousands of degrees hot. The Earth suffered mightily as well, with a significant piece of its crust and mantle vaporized and its surface turned into an ocean of magma.

Much of Theia was eventually incorporated into the Earth. Further out in space, however, the Earth became encircled by a mass of rocky debris that had formerly been a part of the two protoplanets' mantles. In a short time (perhaps only a few years), the debris coalesced under the pull of gravity and formed our Moon.

At the time the Moon formed, it was only 15,000 miles away, as compared to 239,000 miles today. At that distance, the Moon would have appeared twenty times larger than it does today and would have reflected hundreds of times more sunlight than at present. At the time, the Earth rotated on its axis every five hours and the Moon orbited the Earth every 84 hours. Things were moving fast.

The Earth and the Moon pulled on each other with their gravity, causing the rotation of each to slow. Over the ages, the tug of the Earth has caused the Moon to stop rotating all together. The Moon is still working on the Earth, though it has lengthened our days from 5 to 24 hours. Eventually, the Earth too will stop spinning, and Moon and Earth will be tidally locked, each showing the same face to the other for the remainder of their days.

As the Earth's rotation slowed, the rotational energy was transferred to the Moon, causing an increase in its orbital speed, which has caused the distance of its orbit to increase to its present size. The Moon is continuing to recede from the Earth at the leisurely rate of an inch and a half per year. The good news is that the Moon's recession will eventually stop, though that will take 50 billion years or so. The bad news is that the red giant Sun will have already swallowed up Earth and Moon many billions of years before that could happen.

I'm sure some you are saying to yourselves, "Okay, the Earth has a big moon. But how does that affect habitability?" Good question. There are a couple of ways.

First, Earth's equatorial plane is tilted at about 23.5 degrees with respect to the Sun, which itself was the result of the collision with Theia. This axial tilt gives us our seasons. The tilt varies by about a degree over the course of about 40,000 years (from 22.1 to 24.5 degrees). One degree might not sound like much, but it causes huge climactic shifts and is partially responsible for the ice ages. The Moon's gravitational influence stabilizes the Earth's tilt. Without a large moon like ours, the variance in Earth's tilt would be as great as ten degrees rather than only one. Ice ages then would be ten times as severe, with glaciers covering all or nearly all of the Earth.

I note that human beings had no real civilization until the last installment of the ice ages ended 11,700 years ago. The ice is due to return at some time between tomorrow and 10,000 years from now. It is interesting to contemplate the effect it will have on our civilization. Now think of it as being ten times worse.

Another effect of having a big moon has to do with the Earth's speed of rotation. 4.5 billion years ago, as mentioned previously, the Earth rotated every 5 hours. Without the Moon, the day would now be only slightly longer, say six to eight hours. For a day to be that short, the planet has to rotate at extreme speeds. Despite their great size, Jupiter and Saturn have 10-hour days. The rapid rotation required for such short days, and the resulting friction between their surfaces and atmospheres pulls the air into narrow, streaming belts of wind reaching speeds of up to 300 miles per hour and creates hurricanes that can persist for years or even centuries. Without our large Moon, winds would be so fast and powerful that plants would have to be short, deeply rooted and ground-hugging while flying creatures would have to be, well, there wouldn't be any. Life in a perpetual hurricane would be significantly limited.

So, what are the odds that a planet will have such a large moon? Well, the problem with having a big moon is not so much in getting one but in keeping it. In the early phases of planet formation, the protoplanets are bombarded with debris. If a moon like ours forms while the bombardment is ongoing, it would likely be destroyed or knocked out of orbit by these collisions. Thus, to retain a large moon, the moon must form toward the end of the violent formation period. A recent study modeling the early solar system showed the formation of 180 planets. Only fifteen (8%) of the planets were able to hold on to large Moon-like moons. Not conclusive, but informative.

Now, back to the math. As we have seen, it is not a given that there will be a habitable planet suitable for the development of a technological civilization in all stable solar systems. Not by a long shot. A planet must almost certainly be in the liquid water habitable zone and remain in the zone for a lengthy enough time to give life a chance to develop. At the same time, it must also be within the UV habitable zone. It must be the right size to have a livable atmosphere, and it likely must have a magnetic field, plate tectonics and a large moon to create the conditions necessary to sustain life. Since these factors likely range from uncommon to rare, though recognizing the speculative nature of this exercise, I would have to think that no more than one percent of the stable solar systems could reasonably be expected to have a habitable planet. This leaves us with one hundred thousand planets (i.e., 200 Billion [stars in the galaxy] x 5% [GHZ stars] x 20% [G and K stars] x 50% [non-binary stars] x 1.0% [stable planetary systems] x 1.0% [habitable planets] = 100,000).

Although we have eliminated a huge number of planets, we still have a solid hundred thousand candidates capable of sustaining life and technological societies. With so many civilizations broadcasting, communicating and blasting out electromagnetic transmissions, it would seem that SETI's cosmic telephone would be ringing off its hook. But it's not. What's the problem? Well, there is one more matter we must consider.

7. It's About Time

[For purposes of this section, we will consider a planet to be technologically habitable if it has all the ingredients necessary to harbor intelligent, technological civilizations at some point in its existence (to be referred to as "Technologically Habitable Planets" or "THPs").]

It would take a really good camera, but, if we could somehow take a single photo of all 100,000 of our technologically habitable planets, some of the planets would be very old, some newborns, most somewhere in between. Just because a planet is capable of harboring a technological civilization, however, doesn't mean it is doing so right now. Some of the planets in the photo may have active civilizations. Others may be in their equivalent of the age of dinosaurs, still millions of years away from intelligent life. Still others may have had civilizations that thrived hundreds of millions of years ago but have now vanished as utterly as Melville's Pequot Indians. Indeed, if the photo had been taken a couple of hundred years ago, the Earth would be grouped with the not yet technical.

If we want to determine the probability of one or more technological civilizations presently existing on a THP today, we will have to do a little math. First, we must multiply (a) the total number of THPs (i.e.,100,000) by (b) the number of years, on average, we can reasonably expect a technological civilization to remain technological, then divide the product by (c) the total number of years we can expect the THPs to remain technologically habitable. By way of example, if we have 50 light bulbs that will switch on at random (one time only) for 10 consecutive minutes over the course of a 100-minute period, and we want to know, on average, how many lights we can expect to be on at any one moment during that 100-minute period, we would multiply (a) 50 (light bulbs) x (b) 10 (minutes a light is on) = 500 (total minutes all lights are on), then divide that number by (c) 100 (minutes in the entire period), and find that, at any one time, on average, there should be five lights on (i.e., 50 x 10 = 500 ÷ 100 = 5). So, let's give it a try.

In attempting to determine how long a technological civilization will remain technological (part (b) of our equation), we enter the realm of rank speculation. But let's do our best. Our species has existed for about 200,000 years. During that time, we have survived war, famine, pestilence and plague, though sometimes just barely. In the future, we should be better equipped to cope with such threats, but then, we are now much more closely linked, making war or disease more difficult to contain and more likely to have catastrophic effects. Then too, we have just recently acquired technological competence. Any co-exiting civilization would likely be far, far older and equally further advanced. If so, would such a civilization have any interest in making contact with us? I doubt that ourselves of 10,000 years from now would have any more to chat about with ourselves of today than we would have with Lucy and her Australopithecines of three million years ago. They might be out there, but would we even know? We knock, but they turn off the lights and pretend not to be home. Or perhaps, after a few tens of thousands of years, a civilization will have grown weary of cavorting around the universe and have settled down to the cosmic equivalent of a cozy fire and a nice cup of tea, not bothering even to look out the window anymore.

Yet, we need a number. At the rate of our technological advancement in just the last few decades, it is hard for me to imagine what our civilization might look like a hundred years from now, much less a thousand or ten thousand. But then, I may not be very imaginative. So, I will give us half again our total years as a species so far. I feel it highly probable that in 100,000 years we will either be a distant memory or in such an advanced technological state that it simply could not be reconciled with our present understanding of technological. We will apply this estimate to all other civilizations.

Next, we must consider part (c) of our equation, that is, the total period during which the 100,000 THPs will remain technologically habitable. In other words, the time from when the first of these 100,000 worlds became technologically habitable to the time when the last becomes technologically uninhabitable.

Generally, planets will exist as long as their stars stay on the main sequence, that would be about 10 billion years for yellow stars and 15 billion for orange. They will be technologically habitable, however, for a much shorter period. For example, the Earth's interior has cooled over time and will continue to do so. At some point, the interior will be too cool to support geologic activity, meaning an end to plate tectonics and to our protective magnetic field. Some studies forecast that this will take place in as early as a billion and a half years from now, when the Earth is about 6 billion years old. Even before then, the brightening sun will have bumped Earth out of the liquid water habitable zone and the UV zone as well. If the Earth is a representative example, we should expect that most THPs will not be able to sustain technological life after they are much more than 6 billion years old.

The universe is about 14 billion years old (13.8, to be precise). In the beginning, as we have discussed, there was only hydrogen and helium. The heavier elements were forged in the interiors of the early stars. It took generations of stars to produce enough heavy elements to permit planet formation. Thus, planets were probably not common until the universe was about 8 billion years old. We'll use that number for present purposes. If so, the oldest of our 100,000 THPs are now about 6 billion years old. Our newborn of THPs will be habitable until about 6 billion years from now. Thus, from the birth of our first THP until our last becomes technologically uninhabitable is a period of about 12 billion years (6 billion years ago to 6 billion years from now). A technological civilization, however, takes time to develop. Our planet is 4.5 billion years old, but we have been technological for less than 200 years. Could it have happened sooner? Perhaps. I think it safe to conclude, though, that it would take a minimum of 2 billion years for a technological civilization to emerge on any planet. That would mean the first technological civilization would have arisen when the universe was about 10 billion years of age and the last at 20 billion, leaving us with a period of 10 billion years during which at least one of the 100,000 THPs on our photo would be able to support a technological civilization.

Using our formula, we now multiply (a) 100,000 (the number to THPs) x (b) 100,000 (the number of years a technological civilization will, on average, remain technological), and then we divide the product by (c) 10 billion (the total number of years that one or more of the THPs will remain technologically habitable). That should give us the average number of technological civilizations that should be in existence at any one time. Interestingly, the answer is one. (100,000 x 100,000 = 10 billion ÷ 10 billion = 1).

So, in our galaxy, there should be, on average, one, but only one, technological civilization at any one time. And here we are.

Before moving on, I would like to preemptively address an objection to these computations that I'm sure at least some of you might have. I recognize that the figures I have used range from established science to rough estimates to barely educated guesses. If we were to change a number here or there, we might find more technologically habitable planets and, thus, more technological civilizations. Or perhaps fewer. Admittedly though, it is certainly possible that there could be a few more than the 100,000 THPs we have computed. I point out, however, that our figures assume that every one of them will have not only life, but complex, intelligent, technological life. As we will see in Parts II and III of this assessment, that may be our most tenuous assumption yet.

II

LIFE

Given a habitable planet, in an orderly solar system, orbiting a stable star, in a conducive region of the galaxy, can we expect to find life? Complex, multicellular life? That is what we will address in this part.

1.   Genisis

In the Beginning. The Mars-sized, hypothesized protoplanet, Theia, slammed into the proto-Earth 4.567 billion years ago. That is a time so distant that our experience in our daily lives leaves us wholly unprepared to grasp just how distant. So, perhaps another one of those handy, illustrative analogies is in order.

Imagine, if you will, that we can lay time out linearly so that we can travel backward in time. Let's say the present time is right on the edge of the Atlantic coast of the United States in the town of Charleston, South Carolina and that every yard to the west equals 1,000 years of time. In that event, and if you are fortunate enough to live to the ripe old age of 83, your entire life would span a length of 3 inches from the water's edge. Columbus would have undertaken his first voyage of discovery (in 1492 A.D.) at a distance of 19 inches from the coast. Jesus Christ and Julius Caesar would have ministered and fought, respectively, 6 feet away from our starting point. The Sumerians would have invented writing (in 3,000 B.C.) 15 feet away, and our species would have come into existence (200 thousand years ago) a couple of hundred yards down the road. Our next stop will be the KT event, when the asteroid that wiped out the dinosaurs struck the Earth (65 million years ago [65 MYA]). To visit that event, though, we'll have to get into the car and drive all the way to Ridgeville, South Carolina, some 38 miles to the west. Then, we can drive another 90 miles to Augusta, Georgia, just across the state line. There, we will be able to view the emergence of the very first dinosaurs (230 MYA). To witness the collision between the Earth and Theia, however, we would have to drive a bit farther. We would first have to hop back into the car and take Interstate 20 to Atlanta and then to Birmingham, Alabama, where we would exit on to I-22. In Memphis, we would exit on to I-40 and follow it across the states of Arkansas and Oklahoma and the panhandle of Texas. Then on to Albuquerque, New Mexico, Flagstaff, Arizona and finally Barstow, California, where we would head south on I-15 to Los Angeles and the Pacific coast, some 2500 miles from our starting point. You get the idea… the Earth is ancient, and the big thwack was a long, long time ago.

At the time of the collision, the solar system was only about fifty million years old. The Earth and Theia were both still accreting, gobbling up the dust and debris as they circled the Sun in their common orbit. The rule is, however, that two planets cannot occupy the same orbit, or at least not for long. If they do, they will eventually collide. When they do, the bigger one always wins. Earth, at three to ten times the mass of Theia, was, of course, the winner. Theia was completely vaporized. Its heavy, iron-nickel core sunk down through the now molten Earth and merged with the Earth's core. The remaining rock boiled away into a silicate vapor tens of thousands of degrees hot. The Earth did not emerge unscathed, however, with a significant portion of its own crust and mantle vaporized as well. The Earth was hellishly hot, but space is vast and cold. Within a few days or weeks, temperatures had dropped sufficiently for the vapor to condense, and orange hot silicate droplets began to rain down on to the molten surface of our planet.

Higher up in space the remaining vapor cooled and within a few years coalesced to form our Moon. At its formation, the Moon was only 15,000 miles from the Earth. Today it is 239,000 miles away. The close proximity of the two bodies and their gravitational influences caused the Earth/Moon system to spin frenetically. The Earth spun wildly on its axis, a day lasting only 5 hours, while the Moon orbited the Earth every three and half days, causing enormous tides to rip around the molten ocean. The gravitational pull of the Moon on the Earth, however, transferred rotational momentum from the spin of the Earth to the orbit of the Moon, gradually slowing the Earth's rotation and causing the Moon to recede (the Moon is still receding, by the way, at a current rate of 3.82 centimeters or 1.5 inches per year). A billion years after Theia slammed into the proto-Earth, the day had lengthened to 12 hours, then to 18 hours after another billion years.

The Earth continued to cool. In a few thousand years, temperatures had dropped to the point where a thin, hard crust of black basalt appeared, completing the layering of the Earth into the iron-nickel core (an inner core with a radius of 760 miles and an outer core with a radius of about 1,355 miles), the mantle of magma (1800 miles) and the crust (18 miles under the continents and 3 miles under the oceans). Basalt, which flows out of volcanoes as lava, is abundant in the inner solar system, forming most of the crust of Mercury, Venus and Mars, as well as the beds of the oceans here on Earth. Being less dense than the magma of the mantle, it floats. The newly formed crust was punctured by hundreds of volcanoes, which oozed rivers of basaltic lava and belched out volatiles that had been dissolved in the magma. Soon a new atmosphere composed of nitrogen, carbon dioxide and water vapor, along with small amounts of methane and hydrogen sulfide, replaced whatever atmosphere that had been there before the impact with Theia.

The enormous amount of water vapor in the atmosphere would have filled the skies with thick clouds, creating a worldwide steam bath. Once the atmosphere cooled below 212 F, however, the rains came. Torrents lasting perhaps thousands of years would have inundated the Earth, eventually forming a world encompassing ocean. When exactly that occurred, we do not know. The Earth's original crust is long gone, so we have no rocks from the first half billion years that would give testimony to the conditions on Earth at that time. It was long accepted that the oceans were not formed until 3.8 billion years ago. Recently, however, zircon crystals were found in three-billion-year-old sedimentary formations in the Jack Hills of Western Australia. These hard, resilient, sand-sized minerals were formed much earlier than the rock in which they were preserved, analyses indicating that the zircons are as much as 4.4 billion years old. The crystals could only have been formed in a liquid environment, inferring the existence of oceans only a hundred million years or so after the impact. Many, however, remain unconvinced.

In any event, we are confident that the Earth was covered by a deep ocean no later than 3.8 billion years ago. There was then little land to speak of, mostly the tops of volcanoes that had grown tall enough to peak out over the waves. There were also likely to have been a few whitish-gray islands of granite scattered about. The granite formed when the basaltic crust, which completely encased the molten mantle, began to melt in places by the trapped heat of the interior. The melted basalt would have interacted with silicates in the magma to form granite. Granite is about ten percent less dense than basalt and so floats atop the basalt like a block of ice on water, which itself is about ten percent less dense than water. The process of granite formation at that time, however, was very inefficient. Not until plate tectonics got underway about three billion years ago was granite production sufficient to begin forming continents.

As a result of convection currents in the magma of the mantle, the Earth's crust was shattered into seven or eight or twelve major plates (depending on who's doing the counting) and a number of minor plates, which ride upon the ocean of magma below. When two plates meet, one subducts, or slides under the other. The subducted crust is forced down into the mantle, where it melts back into magma. At the same time, new crust is created by upwellings of magma from deep within the mantle along divergent plate boundaries of the mid-oceanic ridge system.

Because granite is less dense than basalt, it floats like a cork and is never subducted. Thus, whereas the oceanic basaltic crust is continuously being replaced, the continental granite lives on forever. Consequently, very little of the basaltic ocean floor is older than 125 million years, while continental granite can be billions of years old. Since new granite is being continuously formed and never goes away, the continents have grown over time and now compose about thirty percent of the Earth's surface. The process is ongoing and will continue to shrink the oceans. Prior to three billion years ago, however, the granite island covered only a marginal portion of the Earth's surface.

A sizable portion of the atmosphere at the time was composed of the greenhouse gases, including water vapor, carbon dioxide (at concentrations hundreds of times higher than today) and a bit of methane. That was fortunate. Four billion years ago the Sun was only about seventy-five percent as energetic as it is at present. If the Sun was only that intense today, the oceans would freeze solid.

The collision with Theia was not the last such event. Not by a long sight. As it orbited, the Earth continued scooping up space debris, which rained down on the planet. Some of the junk could be quite large, 300 miles in diameter or more.

A 300-mile chunk of rock would not have seriously endanger the Earth, which is 8,000 miles in diameter, but it could wreak havoc on the surface. It is estimated that an impact with such a rock could boil away an ocean 10,000 feet deep, scalding the bottom and sterilizing any early life that may have had the temerity of trying to make a living on the young Earth. The barrage continued for some four hundred million years, when suddenly…it got worse. Beginning about 4.1 billion years ago, the Earth and the Moon were mercilessly pummeled by space debris, possibly as a result of an orbital shift between Uranus and Neptune which disrupted a zone of the inner asteroid belt. This Late Heavy Bombardment, as the period is termed, finally came to an end by about 3.8 billion years ago.

And that's the way it was between 3 and 3.8 billion years ago; the bombardment of space debris had largely ceased; a world-wide ocean covered the Earth, dotted here and there with volcanic and granite islands; an atmosphere composed, as it is today, mostly of nitrogen, but with high levels of carbon dioxide and small amounts of methane and hydrogen sulfide (but essentially no oxygen); short days; a large, nearby Moon lighting up the night and generating powerful tides; and a dim Sun adequate to keep the world from freezing over only because of the presence of high levels of greenhouse gasses in the atmosphere. And that's the way it was when life first emerged on Earth.

What is Life? What is water? That was a question asked of learned natural philosophers (scientists, sort of) in the eighteenth century. The responses included:

A substance that is clear and wet.

A fluid that sustains life.

Something that freezes when cold, soaks into wood and flows downhill.

The correct answer, of course, is H_2O, a molecule composed of two atoms of hydrogen and one of oxygen. Such a definition reflects a fundamental understanding of exactly what water is. The "definitions" of the eighteenth century "scientists" were actually only descriptions of the properties of water, some more accurate than others, all incomplete, none terribly helpful.

What is life? If we are going to explore the probabilities of its existence, we will probably want to know what it is. Here's what the experts say:

Life is "any population of entities which has the properties of multiplication, heredity and variation."

Alternatively, life is "an expected, collectively self-organized property of catalytic polymers."

Or, if you prefer, it is "an open system that maintains homeostasis, is composed of cells, has a life cycle, undergoes metabolism, can grow, adapts to the environment, responds to stimuli, reproduces and evolves."

Then there is "a network of processes of production of components such that the components (1) continuously regenerate and realize the network that produces them and (2) constitute the system as a distinguishable unit in the domain in which they exist."

I could go on, but you get the drift. Ask a hundred scientists, and you'll get a hundred different answers, some more accurate than others, all incomplete, none very helpful. As the philosopher, Carol Cleland, and the planetary scientist, Christopher Chyba, have pointed out, as with the eighteenth century definitions of water, these are not definitions at all but merely descriptions. The scientists of the eighteenth century had no knowledge of molecular structures and thus could not define water. Similarly,we simply do not have a fundamental understanding of life at present, and consequently scientists are left to describe rather than define.

Without a definition, we will at least need a good description. Most scientists today believe that life must be a chemical system, that is, a molecular system that undergoes chemical reactions. The chemical system must be able to metabolize, meaning that it must be able to grow and sustain itself by gathering energy and atoms from its environment. It must also be able to reproduce. Reproduction by itself, however, is not enough. Crystals can spontaneously form other identical crystals, but they never change. To be alive, the entity must be able to change. Variation allows for the possibility that some populations will be better able to adapt to environmental change and to become more complex, or, in a word, to evolve. With such factors in mind, NASA's Exobiology Panel has proposed the following "working definition" of life: "Life is a self-sustaining chemical system capable of undergoing Darwinian evolution." Not perfect, perhaps, but it will do for present purposes.

First Life. In 1993, the eminent paleontologist, William Schopf, announced that his team had uncovered the world's oldest fossils. He claimed that he had identified actual single microbial cells preserved in the 3.465-billion-year-old Apex Chert in Western Australia. Moreover, the cells appeared to form filament chains similar to those formed by modern day photosynthetic microbes.

The claim was astounding not only because of the great age of the fossils but because it had been generally accepted that the first photosynthetic microbes did not appear until about 2.4 billion years ago with the first evidence of planet wide oxidation.

It was a bold claim, but Schopf was a respected professor at UCLA and one of the world's leading experts on microfossils. He and his students had already catalogued dozens of microbial species from rocks more than two billion years old, and he had developed a rigorous protocol for examining geologic formations for ancient microbial life. Accordingly, his claims were fully accepted and even regularly included in textbooks on the subject.

Schopf's fossils remained the oldest for just three years. In 1996, paleontologists from the Scripps Institute of Oceanography announced they had found evidence of life in 3.85 billion-year-old rock in the Isua region of Greenland. What they found were not actual fossil cells but rather the mineral apatite, which reportedly contained two different isotopes of carbon with a ratio characteristic of life today. Again, in light of the credentials of the discoverers, these findings were almost universally accepted. Since the purported date of the find coincided with the end of the Late Heavy Bombardment, the discovery was consistent with the belief of many paleontologists that, given the proper environment, life should arise naturally and quickly. Evidence that life appeared almost the day after the bombardment ceased was seen by them as confirmation of their cherished views.

And the hits kept coming. In 1999, scientists with the Australian Geological Survey Organization stunned the paleontology world with a paper published in the journal *Science*, claiming to have found chemical fossils, known as biomarkers, in cores drilled through 2.7 billion-year-old sedimentary rock in the Pilbara region of Western Australia. These biomarkers indicated the presence of oxygen producing cyanobacteria (which we will address in more detail later) as well eukaryotic cellular life (which we will also address), which had previously been believed to have first arisen 1.85 billion years ago. These findings, along with Schopf's filaments, re-wrote the history of life, pushing the appearance of oxygenic photosynthesis and eukaryotic cells back by nearly a billion years.

It has been my experience that people tend to see what they want to see and believe what they want to believe. This holds true for scientists as well, particularly when fame, honors and professional standing are at stake.

In 2000, the Oxford paleontologist, Martin Brasier, undertook a thorough reexamination of William Schopf's discovery, employing new imaging techniques. He found that Schopf's Apex fossils differed significantly from any known microbes. Furthermore, the so-called filaments so reminiscent of the photosynthetic microbes of today, actually appeared in the new images as flat, wide sheets. Also, Brasier's colleagues reexamined the sample site. They found that the chert was formed as a consequence of volcanic fluids at high temperatures, rather than the simple layered formations described by Schopf. In 2002, Brasier published his findings in the journal *Nature*, flinging down the gauntlet. What followed was a heated debate between Brasier and Schopf at a conference of the NASA Astrobiological Society. By most accounts Brasier came out far the better. Then, in 2005, Roger Buick, a professor of geology at the University of Washington, re-dated the Apex Chert to 2.5 billion years, which is very old but a billion years younger than Schopf had supposed. And so Schopf's claim went by the wayside.

Subsequently, geologists performed a rigorous analysis of the Isua Greenland formation from which the supposed 3.85 billion-year-old microfossils were found. They concluded that the rocks were actually an ancient igneous formation that had solidified underground at temperatures in excess of 1000C and could not possibly have contained life.

Next it was the Pilbara biomarkers' turn. In 2005, a major geobiological funding agency, the Agouron Institute, supported the drilling of a new confirmation core. This time no biomarkers were found. Further investigation in 2013 revealed that there were indeed biomarkers in the original core, but it turns out that the stainless-steel saw that had been used to extract the sample had been made stainless (by the manufacturer) by high pressure impregnation of the steel with petroleum products, which petroleum products were the biomarkers found in the core.

By now I'm sure you are saying, "okay, but what's the point?" Well, I guess the point is, three to four billion years ago is a long time ago. Rock from that time is extremely rare, limiting the number of sites that can be searched for early life. Moreover, the rock that remains from that distant time has been squashed, squeezed, twisted, contorted, scalded and compressed, as have any fossils within, making fossil detection next to impossible. And that's not to mention that fossilization of microbes is extremely rare. Finally, there is the problem of contamination. Over the past three billion years, life has accessed and exploited every conceivable environment, including subterranean rock, often making it difficult to determine whether a particular microfossil was the original inhabitant of the rock or a more recent intruder. Consequently, claims to discoveries of life from that ancient period, even by highly credentialed scientists following meticulous protocols, will always be subject to question.

What then can we say about the first life? Well, rock from the early Earth is long gone, so we cannot know whether life existed prior to the end of the Late Heavy Bombardment 3.8 billion years ago. It appears, however, that conditions on the very early Earth (4.5 to 3.8 billion years ago) would not have been conducive to the emergence of life. On the other hand, there are a number of identifiable microfossils of advanced, sophisticated cellular life from as early 2.9 billion years ago. It is reasonable to infer that such life had been around for some time prior to that. But for how long? A few million years? From the end of the bombardment? Or some time in between? Any of these dates is possible or even plausible, but without evidence we simply don't know. So we'll just have to say that life first appeared on this planet between 3.8 and 3 billion years ago and leave it at that.

Abiogenesis. Since the time of Aristotle, it was generally believed that living organisms could arise from nonliving matter through the process of spontaneous generation. Examples could be found everywhere: mold suddenly appearing on bread, maggots on old meat, frogs and salamanders from mud and mice from old rags. The idea held on for a couple of millennia, until Louis Pasteur famously disproved the theory in his 1859 experiment. Pasteur boiled meat broth in a flask topped with a downward curved glass gooseneck that prevented particles from falling in while still allowing the free flow of air. After removing the heat, he let the broth sit for an extended time. The broth remained free of growth. When he removed the gooseneck, however, the broth quickly became clouded, showing that the organisms had invaded the broth came from the air rather than having been generated by the broth.

And so, the theory of spontaneous generation was jettisoned and replaced by the concept of biogenesis, that is, life comes only from prior life. Of course, if you keep peeling back the layers of the onion there must come a point when there is a first life that, absent divine intervention, must have been spontaneously generated from non-living matter. That process is now referred to as abiogenesis. So, spontaneous generation is an obsolete, foolish idea that has been disproved and dismissed; only it must have actually happened a long time ago but doesn't happen anymore. Got it?

Soup. Scientists have been pondering the emergence of the living from the non-living for a long time. In an 1871 letter to the botanist, Joseph Hooker, Darwin famously speculated that life began in a "warm little pond, with all sorts of ammonia and phosphoric salts, light, heat, electricity, etc. present." In 1922, the Russian chemist, Alexander Oparin, proposed that life arose from a body of water that had over time become enriched with organic molecules. The molecules of this "primordial soup" eventually aggregated and self-organized into a chemical system that could duplicate itself. Though short on specifics, the soup idea took hold.

In a 1951 seminar, the Nobel Prize winning chemist, Harold Urey, proposed that life-triggering organic molecules could have been produced in abundance in Earth's primitive atmosphere of hydrogen, methane and ammonia. This conveniently explained why life could have spontaneously generated in the distant past, but not today. Urey's eager graduate student, Stanley Miller, who was in attendance, thought he might score some points by putting his professor's musings to the test. He did so by partially filling a flask with water, which he linked with glass tubing to another flask filled with the aforesaid gases. He warmed the first flask with a flame to simulate the oceanic evaporation, and he sparked the flask containing the ammonia, hydrogen and methane with an electrode to simulate lightning (somehow avoiding an experiment-ending or worse explosion). Soon the water turned yellow, then a deep red, while black gunk began oozing down the sides of the second flask. An analysis revealed that, just as his professor had predicted, Miller had created a complex mixture of organic molecules, including several amino acids.

The experiment brought Stanly Miller much professional recognition, and fame as well. He had transformed the study of the origins of life from an armchair, academic exercise to a verifiable, experimental science. Soon everyone was making soup. Some scientists varied the recipe, adding such things as hydrogen cyanide, formaldehyde and powdered minerals, while others used other forms of energy, such as ultraviolet radiation. Their experiments yielded all manner of organic molecules, including dozens of amino acids, membrane forming hydrocarbons, metabolic acids and even some of the building blocks of DNA and RNA. These were heady times. Unfortunately, though, there was a fly or two in the soup.

First of all, it was determined that the Earth's early atmosphere was composed not of ammonia, hydrogen and methane, as previously supposed, but of nitrogen and carbon dioxide. You can zap electricity through those gases all day long and not much of interest will happen. Then there's the soup itself. It is non-reactive. It is in a state of equilibrium. It doesn't want to do anything. Open a can of Campbell's chunky beef and vegetable in a sterile environment and watch. You can wait a day, a month, a year, a millennium, and nothing will happen. It'll just sit there like... well, a can of soup. Moreover, electricity and ultraviolet light tend to break down rather than form organics. Proteins, which are formed of long chains of amino acids, will be ripped apart by lightning, as you might imagine, or by the intense ultraviolet radiation that roasted the early Earth. Besides, even if organic molecules were created in abundance, the ocean is vast and the soup would have been far too dilute to bring together all of the organics needed for life. And that's not to mention that no mechanism was ever proposed for assembling the organic building blocks into any sort of self-sustaining, self-replicating form.

Vents. In 1977, scientists descended to a depth of 8,000 feet in the submersible, *Alvin*, to explore a volcanic ridge off the Galapagos Islands. Specifically, they sought to find and explore hydrothermal vents that apparently were the source of warm water plumes that had been detected from the surface. To their amazement, they found an abundance of previously unknown organisms living around the vents. These includes giant clams, albino crabs, tubeworms up to six feet long, eyeless shrimp, semitransparent vent octopi and dozens of species of snails, slugs and other gastropods, all in population densities rivaling those of the rainforests and coral reefs. The vents themselves were towering black chimneys, some as tall as buildings, pumping out scalding hot, mineral-rich fluids produced by cold seawater seeping into the magma below and then spewing back up like a geyser. Because the discharge looked like black smoke, the chimneys were dubbed black smokers.

The crushing pressure of the frigid, black ocean deep, side by side with acidic, superheated vent fluids seemed wholly uninhabitable, and scientists were perplexed by the multitude of life flourishing under such extreme conditions. They soon learned that hydrogen sulfide was a significant component of the smoke. Sulphur bacteria living at the interface between the vents and the ocean were able to extract the hydrogen and attach it to carbon dioxide to form organic matter.

Before long, some 300 vent fields had been identified along the oceanic ridges of the Pacific, Atlantic and Indian Oceans, sheltering some 750 species. It had always been assumed that life was powered by sunlight and that life, therefore, must have arisen at the surface. The discovery of a thriving, diverse, world-spanning ecosystem that ran on chemical energy and had never even heard of the Sun forced origins of life scientists to reconsider their assumptions. Was it possible that life arose in the ocean deep and only later colonized surface waters? Indeed, a deep origin solved many of the difficulties associated with a surface origin. Deep life would have been protected from ultraviolet radiation and bombardment by space debris. Additionally, the Earth was a water world in its early days, providing no place for organic molecules to gather and concentrate. The ocean ridges, on the other hand, had rocks and mineral surfaces where organics could aggregate. Most significantly, the interface between the vents and the ocean were in a state of dynamic disequilibrium and highly reactive, permitting the kinds of chemical reactions necessary for life.

Gradually, the idea of a deep origin of life gained support. The theory, however, was sternly resisted by the surface origin crowd, led by Stanley Miller, who felt that their long-held turf was being wrongfully invaded by unsupported speculation. In an interview with *Discover Magazine*, Miller sniffed: "The vent hypothesis is a real loser. I don't understand why we even have to bother to discuss it," illustrating again that scientists, like everyone else, are not strictly governed by reason. They did have their points though. Hydrothermal temperatures at the vents were over 400 degrees C (750 F), which is far too hot for amino acids and other essential molecules to exist. Moreover, life on the vents today is dependent on the presence of oxygen, which was not present in the early oceans. In spite of these and other problems, the deep origin theory has become generally accepted.

Emergence. Fundamentally, there are two processes required for the formation of life, metabolism and genetics. Metabolism is the ability to manufacture biological structures from matter in the environment using a source of energy. Genetics is a blueprint for the construction of the organism that can be transferred from one generation to the next. Genetics and metabolism, however, are two distinct and dissimilar chemical systems, making it highly improbable that they emerged simultaneously. More likely, one or the other system appeared first, and that system eventually developed the other. Thus, very early on, origin of life scientists split into the "genetics first" and the "metabolism first" camps.

The genetics fist camp views metabolism without genetics as simply a chemical reaction. When it's over, it's over. Life, to them, is all about self-replication, passing genetic information from one generation to the next, and evolution.

Today, replication is controlled by DNA, which are long chains of nucleotides. No one suggests that a complex molecule like DNA could have been produced randomly. Moreover, there is a chicken and egg issue with DNA formation. DNA codes for the production of proteins. Proteins, which are long strings of amino acids that are very different chemically from DNA, perform most cellular functions. RNA, another, though far simpler, string of nucleotides, transfers the genetic information from DNA to cellular structures known as ribosomes, where proteins are formed. One class of proteins, known as enzymes, speed up chemical processes that would otherwise occur too slowly to be of use to life. One of these chemical processes is the formation of DNA. Thus, DNA contains the information that is necessary for the formation of proteins, yet that information cannot be retrieved or copied without the assistance of protein enzymes, which have to be made by DNA.

A way out of this dilemma appeared to have been found in 1982 when it was discovered that RNA could not only carry genetic information but could also serve as a catalyst to replicate itself. According to this RNA world hypothesis, life began with the appearance of the first RNA molecule, which acted as a catalyst for its own self-replication from a nucleotide soup. In this way, the self-replicating RNA performed the functions carried out today by DNA, RNA and proteins.

The RNA world, however, had its problems. First, no examples of self-replicating systems of RNA are found in nature. Moreover, it is difficult to conceive how a long RNA molecule could be formed initially without the assistance of enzymes, which did not then exist. As one researcher put it, "[t]he spontaneous appearance of RNA chains on the lifeless Earth would have been a near miracle."

Given these difficulties, it has been suggested that there may have been a pre-RNA world, when there existed a chemically simpler RNA-like molecule that possessed both information storage and catalytic properties. The only problem is that there are no such substances in present-day cells, and, if there ever had been, they would have left no fossil record. And so most origins of life scientists have turned to metabolism first theories.

There are any number of metabolism first hypotheses. Perhaps the most innovative was proposed by a chemist/patent attorney with the excellent name of Gunter Wachtershauser. His so-called Iron-Sulfur world theory is very much a back-to-basics model. Wachtershauser first postulates that early proto life was autotrophic rather than heterotrophic, meaning that it manufactured the essential biomolecules on site instead of scavenging them from the prebiotic soup, as had been generally assumed. It might seem easier to simply use the biomolecules in the environment that had already been made by other chemical processes than to invent a mechanism for constructing what was already there and available. The soup, however, would not only have been hopelessly dilute, it would have been a random mixture of organic molecules, varying from place to place and which might or might not contain all of the molecules needed by the emerging life. Manufacturing the required molecules in a simple, regular pattern a few atoms at a time seems a more straightforward approach than waiting on the serendipitous conjoining of the necessary ingredients.

Manufacturing anything, of course, requires a steady flow of energy and building blocks. For these, Wachtershauser turns to the hydrothermal vents. In the volcanically active mid-oceanic ridges, seawater percolates down thorough the ocean floor into the magma, where it is heated to hundreds of degrees and saturated with dissolved metals and sulfides. The superheated water then blasts back to the ocean above, where it abruptly cools, precipitating an abundance of iron sulfide (FeS) particles. Iron sulfide, being highly unstable, will over time decompose into more stable chemicals. In one reaction, iron sulfide combines with hydrogen sulfide, a volcanic gas common in the vents, to produce pyrite (FeS_2) (a/k/a fool's gold) plus hydrogen gas (H_2), along with a bit of energy. The energy can then drive the hydrogen gas to react with the CO_2 that is dissolved in the seawater to synthesize organic molecules. Wachtershauser envisioned these reactions occurring on the surface of the pyrite, where organics could be catalyzed by iron-sulfur mineral clusters similar to those still found in many enzymes today.

Life, Wachtershauser realized, requires more than random chemical reactions. It must have a metabolic cycle, that is, a series of linked biochemical steps to convert molecules into different, more usable forms. All life today shares a common core of metabolic reactions, at the heart of which is a cycle of simple chemical reactions known as the Krebs, or citric acid cycle. You may remember the Krebs cycle from high school biology. Or maybe not. Anyway, the Krebs cycle normally breaks down organic molecules to produce hydrogen, which is burned with oxygen in respiration, and carbon dioxide, which is expelled as waste. When run in reverse, it does just the opposite, joining carbon dioxide with hydrogen to form organic molecules. The reverse cycle starts with a 2-carbon molecule known as acetate. The addition of one molecule of carbon dioxide builds the 3-carbon pyruvate. Add another, and the 4-carbon oxaloacetate is formed. When the cycle reaches the 6-carbon citric acid, the molecule splits into one 2-carbon acetate molecule and one 4-carbon oxaloacetate molecule, which are the starting points for two new cycles. Thus, the cycle is self-replicating, doubling on each turn. Moreover, each organic molecule created in the reverse Krebs cycle serves as a starting point for other metabolic cycles that synthesize even more complex molecules. And on it goes.

The Iron-Sulfur world hypothesis was truly groundbreaking, leading one scientist to remark that reading Wachtershauser's paper was like stumbling across a scientific paper that had fallen through a time warp from the end of the twenty-first century. Like just about every other idea on the origins of life, however, there were problems with the theory.

For starters, getting hydrogen sulfide to react with carbon dioxide is very difficult, especially at high temperatures. Wachtershauser relied on iron-sulfur minerals to catalyze the formation of organic molecules. Although the iron-sulfur minerals are good catalysts, Wachtershauser's own experiments showed that it does not work. Instead, he was forced to substitute the more reactive carbon monoxide for carbon dioxide to produce organic molecules. There is, however, very little carbon monoxide in the black smoker vents.

Another problem is the high temperatures in the vents, between 250 – 400 C (482 – 752 F). Such temperatures prevent organic synthesis. Finally, organic chemistry doesn't work well on flat mineral surfaces, or so says Nick Lane, the biochemist and popularizer of origins of life topics. In his words, "Organics [on the surface of minerals] either remain bound to the surface, in which case everything eventually gums up, or they dissociate, in which case they are flushed out into the open ocean with unseemly haste."

Geochemist, Mike Russell, and microbiologist, Bill Martin, think that the problems with the Iron-Sulfur world hypothesis stem from Wachtershauser's selection of the wrong kind of hydrothermal vent. Russell and Martin agreed with Wachtershauser that the first life was likely autotropic, synthesizing organic molecule from hydrogen and carbon dioxide. They also agreed on the importance of iron-sulfur minerals as catalysts and the reverse Krebs cycle as the initial means of synthesizing organic molecules. But the black smokers, they believe, were far too hot, violent and dramatic for the gentle chemistry of life to take place. There is, however, a second type of hydrothermal vent. These vents are not volcanic and do not interact with magma. Instead, they are formed along the spreading zones of the mid oceanic ridge when seawater seeps down into the mantle to hydrate mineral rich rocks, such as olivine. Their reaction generates heat and copious amounts of hydrogen gas. The warm, alkaline, hydrogen-rich fluids then rise back to the sea floor. There they cool, react with the salts in the seawater and precipitate out into large vents on the ocean floor.

These "alkaline" vents are not single channeled chimneys venting directly into the ocean like the black smokers. They are, rather, like a mineralized sponge formed of a labyrinth of interconnected micropores. The early oceans were full of dissolved iron, which, according to Russell and Martin, would have precipitated within the vents as iron sulfides, forming catalytic clusters similar to those found in enzymes driving carbon metabolism today. Moreover, carbon dioxide levels then were 100 to 1,000 times greater than today, providing an excellent source of carbon. Additionally, the micropores would have provided a containment for the accumulation of high concentrations of organic molecules. Thus, the story goes, the warm carbon dioxide and hydrogen-rich fluids would have slowly flowed up from the mantle through a labyrinth of catalytic-walled micropores where organic molecules could have been produced, concentrated and retained, while wastes were ventilated away. In this way, the tiny mineralized micropores became the first living cells while the porous rocks of the alkaline vents became the hatcheries of life. Eventually membranes were formed in the pores, and the cells escaped from the vents into the empty world.

The Russell/Martin hypothesis addresses many of the weaknesses of the Iron-Sulfur world, but the hydrogen/carbon dioxide reaction problem remains. A promising theory, regardless.

Another theory addresses the significant problem that RNA is a large and complicated molecule, making it very fragile and easily destroyed. Indeed, water will break up the strings of nucleic acids that make up RNA. As a result, the authors argue, RNA simply could not have been formed in a liquid environment. What is required is dry land and a mechanism to concentrate dilute solutions of the nucleotides that are the building blocks of nucleic acids. They suggest that the ideal location would be a desert through which runs transient waterways. Nucleotides, the theory goes, can be made from common desert minerals. Occasional precipitation would dissolve the minerals and wash them into the resulting waterways. The dilute mineral solution would eventually be deposited into basins, where it would decant and distill, causing the accumulation and concentration of the minerals necessary for the building up of nucleotides.

The problem with this scenario is that, at the time that life emerged, there was precious little dry land and almost certainly no deserts or transient waterways in our water world. There was one place, though, where such conditions did exist.

That was the planet Mars. Mars at that time was largely desert, though it appears to have had numerous lakes and even small seas. Moreover, the mantle contained methane, hydrogen and other gases needed for prebiotic synthesis of the carbon compounds needed for life. Thus, they conclude, life might very well have emerged on Mars.

So, what relevance does possible Martian life have for us? Well, Mars is only about half the diameter of Earth and has just 37.5% of the gravity. Even a relatively small asteroid or bit of space debris could eject a good deal of rock into space, some of which might find its way to Earth. In fact, it is estimated that over the last four and half billion years more than a billion tons of Martian rock has made it to Earth in exactly this way. Any microbes within the rocks could quite easily have hitched a ride and quickly populated their new home, making us all Martians. Earth rocks might also have reached Mars in this fashion, but, for microbes, travel between the two planets was a one-way street. Earth's greater mass and stronger gravity would require much more energy to launch material into space, likely melting and sterilizing the rock in the process.

This "Out of Mars" theory is one more iteration of a much older theory with the disagreeable name of "Panspermia." According to its proponents, which included such illustrious figures as the astronomer, Sir Fredrick Hoyle, life began elsewhere in the universe, but, as a result of some planetary collision, microbes in their resting stages were blasted out into space, where they traveled about until they were deposited on to a suitable planet, which they then colonize. That may be an explanation of how life got started on Earth, but it is not an explanation of how life got started in the first place. It seems to me unlikely that microbes could endure the harsh environment of space for more than, perhaps, a short jaunt from Mars to

Earth. Anyway, if life cannot be generated in Earth's ideal habitat, it must be a rare commodity indeed.

There are other theories: the PAH world, the Protenoid world, the Theioester world, aerosol life, flat life, clay life, to name a few. Each emphasizes one scientific discipline or another, usually the field of specialization of the scientist propounding the hypothesis. As a result, the theories are all over the board. There does not yet appear to be much of a narrowing of the options or any consensus as to the way forward. This leads me to believe that, not only do we not know how life originated, we have no prospect of knowing in the near future.

Easy or Hard. So, is life easy or hard? Is it probable, expected, inevitable? Or is it unlikely, rare, unique? Since we don't really know how life began or when, and we don't even know exactly what life is, that is a difficult question. The easy livers will tell you that life is nothing more than a chemical reaction, or a series of them. Thus, whenever the necessary ingredients are present and the conditions are right, the chemistry will happen. Life, they say, should be easy and abundant. On the other hand, the hard lifers will point out that, since we don't know what ingredients are needed or what the right conditions might be or how the reactions would take place, the easy livers are indulging in wishful thinking. Who is right? Without more, I am afraid, we are in no position to judge.

Despite our ignorance, I think some fair inferences can be made. If, for example, we eventually uncover solid evidence of life 3.8 billion years ago (right at the end of the Late Heavy Bombardment), as the scientists at the Scripps Institute thought they had found, meaning that life started here at the very moment conditions permitted, the only reasonable conclusion would be that life is easy. If, on the other hand, we were to determine that life did not begin until 3.3 billion years ago, that is, half a billion years after conditions on the planet became amenable to life, that would seem to introduce an element of chance into the equation. Perhaps it takes more than simply bringing together the right ingredients under the right circumstances. Maybe a confluence of unlikely events is required, like drawing into an inside straight. We will have to await further discoveries.

There is another matter that deeply perplexes me, though it doesn't seem to have caused the origins of life community much concern, that being the utter absence of diversity of life on this planet. I know some of you are saying to yourselves "what the heck is he thinking about? From parakeets to blue whales, bacteria to carrots, house flies to redwoods, eagles to slime mold, how could there be any more diversity than we already have?" But that's not what I'm talking about. Every living creature on Earth, no matter its shape or size or lifestyle, is the same DNA based life form. That is because we are all related. On this planet, there is a single tree of life (or bush or shrub or whatever the fashionable term is these days). Go back in time far enough and you will find that all living creatures are the descendants of one, single living thing. The great (to some great power) grandmother of us all. No one disputes this. Indeed, they have given her a name, LUCA (the last universal common ancestor).

A success story if there ever was one. Three or four billion years after her time, LUCA's grandchildren by the trillions occupy every nook and cranny of this planet. I sometimes wonder what would have happened if some errant, falling rock had crushed her just before she first gave birth (actually, divided). Would that have been it for life on this planet? Would the world today be void of living creatures? I am sort of being facetious, but sort of not.

I think you get the picture. Why is there only one kind of life on this obviously ideal habitat for life? I would be much more comfortable with the idea that the emergence of life is natural, expected and easy if there were hundreds, thousands or even just a few living creatures on this planet that sprang from a completely separate genesis with an entirely different biochemistry. That would tell me that, at some distant time, life on Earth was sprouting up all over and that the same could be expected to occur on any other habitable planet. That, however, is not what we see.

There are, of course, proposed explanations. For instance, there may have been in ancient times numerous other life forms that have all died off (which raises its own set of questions). Or maybe there is only one way to do life. There were, perhaps, numerous genesis events, all spawning life with the same chemistry, which eventually merged into a single population. Or maybe it was a race. There might have been all manner of proto life, all gradually becoming more complex, all moving toward life. But as soon as one stumbled across the formula that allowed it to metabolize, multiply and evolve, it would have had the world to itself. With no natural enemies and no competition, the first life would have multiplied exponentially, monopolizing resources and crowding out all of the "not quite life."

These explanations have merit, but they are entirely speculative. I think it important to keep in mind that at present there is evidence of only one type of life, though that could change. Until we find an example of another life form, however, we cannot completely rule out the possibility that our tree of life is unique.

Assessment. In Part I of this assessment, we sought out habitable environments. In Part II, we explore whether a habitable planet will be inhabited just because it is habitable or is something more required. In this first section of Part II, we have asked the first question that needs to be asked. Is life common or rare? In view of the present state of our knowledge, we must conclude that we can reach no conclusion. So, we will move on to other matters.

2. Lottery Tickets

I read the other day that the chances of winning the Mega Millions Lottery, as of that date, were about 1 in 259 million and that of the Powerball Lottery 1 in 292 million. The chances of winning both jumped to 1 in 77 quadrillion. Those are slim odds, in my estimation.

Photosynthesis. Oxygen is ferociously reactive. Because of its chemical composition, it wants to combine with almost anything. Its reactivity makes free oxygen a rare commodity indeed. Oxygen is poisonous, combustible and deadly. These dangerous qualities, however, make it a source of incredible power when properly harnessed.

Bacteria can get by on meager sources of energy, such as fermentation, but they are tiny and relatively simple. Oxygen respiration, on the other hand, enables cells to extract comparatively vast amounts of energy by burning food with oxygen. The power of oxygen enables cells to grow large and complex and to become active and energized. Without oxygen, there would be no plants, no animals or any multicellular life at all on this planet. It would be a microbial world, as indeed it was before there was any free oxygen.

Oxygen is critical for complex life for other reasons as well. High up in the stratosphere, ultraviolet rays split oxygen molecules (O_2) in two. The free, single oxygen atoms then combine with other O_2 molecules to form ozone (O_3). Ozone is adept at absorbing UV rays, and the thin layer of ozone that surrounds the Earth forms a shield that protects the planet from UV radiation. That is a good thing, since ultraviolet rays rip apart organic molecules. Without ozone, life on the surface of this planet would not be possible.

Ultraviolet radiation can also split water molecules into its constituent parts, hydrogen and oxygen. Without atmospheric oxygen, the oxygen freed from the split will quickly react with iron in rocks, forming a thin layer of rust. The free hydrogen, being too light to be held by the planet's gravity, will drift off into space. Gradually but inexorably, the oceans will boil away, leaving a barren, lifeless, rusty-red desert. Mars once had, if not oceans, seas and lakes. It likely lost them by this very process. Atmospheric oxygen, however, saves the day. Not only does it produce the ozone shield, any hydrogen that is split from water will react with the free oxygen in the atmosphere, once again forming water, which rains back down into the oceans.

As we have noted previously, the atmosphere of the early Earth contained essentially no oxygen. Today oxygen comprises about 21 percent of the atmosphere. So where did it all come from? Of course, any school child can tell you; photosynthesis, the magical process whereby green plants combine water, carbon dioxide and a drop of golden sunshine to form sweet sugar and a whiff of fresh oxygen, the formula being $6H_2O + 6CO_2 + energy = C_6H_{12}O_6$ (i.e., glucose) + $6O_2$.

And how exactly does it do that? Well, the process is actually quite complex. It involves a 5-step process utilizing two separate light reactions, referred to as photosystems, that strips electrons from water and joins them to carbon dioxide. The reason for the complexity is that water is a very stable molecule that is reluctant to give up its electrons. Carbon dioxide is very stable as well and has no wish to be stuffed with additional electrons. As a result, it takes a lot of energy to make it all work.

To visualize the process, think of the capital letter N. Now imagine that one of the two photosystems is located at the base of the first vertical and the second at the base of the other. It all begins in the first photosystem, when an enzyme referred to as the "water-splitting complex" seizes a molecule of water and pulls electrons out one by one. Being of no further use to the plant, the oxygen is released as a waste product. The green light-harvesting pigment, chlorophyll, then captures a photon of light, which blasts an electron to a higher energy level, like shooting it up to the top of the first vertical of the N. In step three, the electron cascades down the diagonal, giving up its extra energy as it does. That energy is used to make ATP (called the "energy currency of life"), which is used to power the work of the cell. At the base of the second vertical, the electron enters the second photosystem. There, another molecule of chlorophyll captures a second photon, which blasts the electron back up to the higher energy level. At the top of the second vertical, the energized electron is transferred to a molecule of carbon dioxide as the final step of this process and the first step in the sugar making. process, known as the Calvin Cycle.

In green plants and algae, photosynthesis takes place in the chloroplasts. Chloroplasts are tiny organelles within the cells that look all the world like bacteria. That's because chloroplasts are all descendants of a single free-living bacteria that was once long ago consumed, survived digestion and proved highly beneficial to the host. This host cell and its symbiotic partner are the progenitors of all plants and algae.

Since there is only one kind of bacteria that can perform "oxygenic" photosynthesis, we know that the engulfed bacterium was a cyanobacteria (formerly known as blue-green algae). Indeed, the chloroplasts have their own separate ring of bacterial DNA, the sequences of which are almost exactly the same as that of the cyanobacteria, and all plants, algae and cyanobacteria utilize essentially the identical five step process in photosynthesis. Thus, it is clear that "oxygenic" photosynthesis was invented by and has been performed by only the cyanobacteria and their captured descendants, the chloroplasts.

Cyanobacteria, however, are not the only bacteria that carry out photosynthesis. Purple and green bacteria (both the sulfur and non-sulfur varieties), for instance, have been photosynthesizing for a lot longer than the cyanobacteria, perhaps by as much as a billion years. There are major differences, however, between the kind of photosynthesis utilized by cyanobacteria and that practiced by the other bacteria. First, these other bacteria utilize one or the other of the two photosystems but never both. Moreover, none attempt the difficult task of splitting water, opting instead for other sources of electrons, such as hydrogen sulfide, which are comparatively easy to split. And finally, none produce oxygen, giving these forms of photosynthesis the name "anoxygenic" photosynthesis.

It appears that the second type of photosystem (the one that builds sugars) evolved first and that the other (which produces energy only) later. It is hypothesized that an ancestor of the cyanobacteria somehow acquired both types of photosystems, possibly by lateral gene transfer that bacteria so often utilize. The ancestor did not, however, use both systems at the same time. More likely, when building materials were present in its environment, it utilized the second photosystem to produce sugars. Conversely, when building materials were not available, the ancestor used the other system (that produces only energy), which did not enable growth or replication but did allow for survival until better times.

The acquisition of both photosystems did not, in and of itself, lead to oxygenic photosynthesis. There was still the problem of stripping electrons from highly uncooperative water. That problem was solved when the cyanobacteria's ancestor acquired the water-splitting complex that is step one of the 5-step process of photosynthesis. The complex is a tiny enzyme composed of four manganese atoms and one calcium atom in a lattice of oxygen atoms, all wrapped up in a protein. This enzyme catalyzes the stripping of electrons from water, making oxygenic photosynthesis possible. It appears that the evolution of the water-splitting complex resulted from the presence of significant quantities of manganese-containing minerals in the early oceans. The minerals were assimilated by cyanobacteria and somehow incorporated into the active center of the complex. One noted authority noted that there is a huge sense of the accidental about the whole process. But accidents happen and apparently did happen, and the world would never be the same again.

So, when did this all happen? Happily, this is one of those occasions that we actually can date an event in the distant past with some degree of certainty, not from the fossil record, mind you, but from the geologic. As noted previously, oxygen is highly reactive. When the cyanobacteria first began unleashing oxygen into the environment, it did not accumulate. Instead, it reacted and combined with minerals. It wasn't until everything that could be oxidized was oxidized that free oxygen began to accumulate. One of the first minerals oxygen would have encountered was iron. Iron readily dissolves in water, and the early oceans were saturated with it. When oxygen is present, however, the iron rusts and comes out of solution. That is precisely what happened when oxygenic photosynthesis first got going. Vast quantities of iron rusted out of the oceans and accumulated on the ocean beds. These iron oxide deposits now show up in the geologic record as rust-red bands which comprise about 85 percent of the worlds iron ore deposits. These banded iron formations can be dated to 2.4 billion years ago, during a time termed by paleontologists as the Great Oxidation Event.

As an aside, the oxygenation of the world had another consequence not directly related to our inquiry but of interest regardless. As oxygen began combining with existing minerals, the oxidation created many entirely new minerals that had never existed on Earth before. In fact, of the 5,000 known mineral species on this planet, about 3,500 resulted from the oxygenation event. These minerals are not found anywhere else in our solar system.

The takeaway from all of this is that no organism larger or more complex than bacteria could ever have evolved without the presence of free oxygen. The sole source of free oxygen on this planet is oxygenic photosynthesis, and oxygenic photosynthesis evolved only once. That's right... once. About 2.4 billion years ago, a single bacterium learned to build a little manganese mineral cluster that, when coupled with its pre-existing ability to use both photosystems, enabled it to conduct photosynthesis using water. Every green plant, algae and cyanobacteria on this planet is the direct descendant of this single bacterium.

Photosynthesis was not, as the "easy livers" believe of the emergence of life, a natural chemical process that was bound to occur as soon as the conditions were right. Instead, it appears to have been an accident. A highly improbable accident. What are the odds that only one of trillions of bacteria over trillions of bacterial generations would one day stumble upon this world-changing process, without which we certainly would not be here today? But, hey, stuff happens.

Ladies and gentlemen, we have a winner.

The Complex Cell . Our understanding of the emergence of complex life has been a bit like putting together a jigsaw puzzle... a lot of disparate parts that seem to form no coherent picture until suddenly they do.

Modern taxonomy, the science of classifying organisms, began with the Swedish botanist, Carolus Linnaeus, in the eighteenth century. Linnaeus grouped organisms according to shared physical characteristics. He gave us the familiar groupings, from smaller to larger, species, genus, family, order, class, phylum and kingdom.

In Linnaeus' system, there were just two kingdoms, animals and plants. This seemed a logical division of the world's organisms, since that is pretty much what we see, and, indeed, we have long divided biology into zoology and botany. Some organisms, such as fungi, algae, protozoans and bacteria, didn't fit very well under either category. Nonetheless they were shoehorned into one or the other of the two kingdoms anyway, largely depending upon whether or not they moved.

As time went on, it became increasingly clear that the two-kingdom arrangement was unworkable. In 1969, R.H. Whittaker proposed a five-kingdom classification system, adding to animals and plants the kingdoms of fungi and protists (a grab bag of single celled organisms). Bacteria, though single celled, were segregated into their own kingdom, Monera.

From early on, it was clear to scientists that bacteria were significantly different from all other life forms. In the first place, they are tiny, measuring only on average about $1/100^{th}$ of the size of the cells of all the other life. Moreover, unlike the cells of everything else, they have no membrane-bound nucleus, only a single circular free-floating chromosome. The absence of a defined nucleus led to bacterial cells being given the name prokaryotes, from the Greek words "pro" (before) and "karyon" (kernel), to distinguish them from all other cells, which were called eukaryotes, from the Greek for "true kernel."

All eukaryotic cells (with limited exceptions) are very similarly constructed and are clearly closely related. In addition to a membrane enclosed nucleus, they contain membrane bound organelles that perform a variety of cellular functions, just as organs do in the human body. They all have rough and smooth endoplasmic reticulum that form an interconnected network of membrane enclosed tubules for the transportation of proteins. All have a dynamic internal skeleton that provides support while permitting the cell to change its shape and to move. What's more, all eukaryotic cells reproduce through mitosis and, now or at some point in the past, sex.

Prokaryotic cells, on the other hand, are starkly simple, with no organelles or much of anything else inside. They reproduce asexually through fission, and they are surrounded by a rigid cell wall that gives them their shape and which, interestingly, has a completely different chemical composition than the cell walls of those eukaryotes having cell walls (plants and fungi).

Bacterial prokaryotic cells and eukaryotic cells do have some traits in common. Both are DNA based, both have cell membranes separating the inside of the cell from the surrounding environment and both are filled with cytoplasm, a gel-like fluid composed of water, salts and protein. They are clearly related, but only remotely so.

Since bacteria and eukarya diverged so long ago, I suppose it may not be surprising that their cells are so different (even though some of the differences are so fundamental that it is difficult to determine how they could have occurred). What is surprising is the complete absence of intermediaries between the two groups. If eukarya arose from small, simple bacteria, there should be a trail of organisms gradually growing larger and gaining complexity. But that is not what we see. Instead, there is a deep chasm between the two groups of organisms.

It is as if there were worms and human beings with nothing in between.

A piece of the puzzle was solved, but then made more complicated, by the work of the groundbreaking microbiologist, Carl Woese. In the 1970s, Woese searched for a means to accurately measure the evolutionary distances between living organisms. Eventually, he settled on ribosomal RNA.

Cells utilize proteins to carry out their cellular functions. Proteins are synthesized in ribosomes, which are membrane-encased organelles within eukarya and floating freely within prokaryotes. The code for protein synthesis lies within the DNA. One type of RNA transcribes the instructions from the DNA, another transports the instructions from the DNA to the ribosomes and a third, ribosomal DNA, supervises the manufacture of the protein within the ribosomes. Thus, ribosomal DNA is essential to the functioning of cells in all creatures, from bacteria to elephants.

Woese then compared the nucleotide sequences of ribosomal RNA from numerous organisms. At one point, Woese analyzed a type of bacteria then known as archaebacteria, which are sometimes referred to as extremophiles because of the extreme environments where many may be found, such as hot springs and salt lakes. To his astonishment, the analysis showed that the archaebacteria were not bacterial at all. Not even close. The archaebacteria (now known as "archaea") and bacteria certainly looked alike, both being prokaryotic and thus tiny, simple and having no membrane bound nucleus or organelles, but they were very different from each other on a fundamental level. They have completely different biochemical make-ups of their cell walls and outer membranes, and they follow different metabolic pathways. Indeed, they were as distinct from bacteria as are plants and animals. In fact, they were much more closely related to eukarya than to bacteria.

In light of this evidence, Woese proclaimed there were not two kingdoms of living things, as established by Linnaeus, or five, as Whitaker had proposed, but three: bacteria, archaea and eukarya (i.e., everything else.) He then pulled up the long-revered evolutionary tree of life by its roots and drew his own tree (or bush) based on genetic relationships rather than morphological (structural) similarities. Woese's tree showed a branching of bacteria and archaea diverging from a common ancestor at the very dawn of life, and a later branching of eukarya from archaea, with plants and animals depicted as merely a twig on the eukarya branch.

Woese's findings were not well received, many scientists being indignant that simple, tiny prokaryotes should be granted two of the three kingdoms, while all visible life was relegated to an offshoot of the eukarya branch of the tree of life. Over time, however, Woese's work has been repeatedly verified and is now generally accepted by the scientific community.

That bacteria and archaea diverged long ago from a common ancestor and eurkarya more recently from archaea explains why bacteria and eukarya are so different in composition. There are, however, still no intermediaries between archaea and eukarya. The vast chasm between the kingdoms remains. Even more perplexing, is the fact that eukarya share certain traits with bacteria that are not shared with archaea. How could that be?

The mystery of the missing intermediaries was eventually solved. As it turns out, there are none because there can be none. Both archaea and bacteria are physically restrained from growing larger. That is because they generate energy from their outer membrane. As we all know, the ratio of the surface area of an object to its volume diminishes as the size of the object increases. Thus, as a prokaryote grows larger and its surface area to volume diminishes, it will produce less energy per unit of volume, meaning that at some point the surface area will be relatively too small to produce sufficient energy to power cellular function. With increased size foreclosed from them, the prokaryote's survival strategy became rapid reproduction. Rapid reproduction requires simplicity. Thus, prokaryotes refrained from developing any complex internal structures, even regularly discarding any genes that are not immediately required by the cells to function in their present environment (though they are able to acquire new genes that might enable them to survive changes in their environment through lateral gene transfers with other prokaryotes). Thus, all prokaryotes, whether bacteria or archaea, are forever limited in size and complexity.

But if prokaryotes can't grow larger, where the heck did eukarya come from? Did they just pop up into existence from nothing? The answer to this question is related to the mode of energy generation utilized by eukarya.

Eukarya do not generate energy from their outer membrane as prokaryotes do. Instead, all eukarya have tiny organelles known as mitochondria that generate energy for cellular function. Mitochondria, often referred to as the powerhouses of the cell, generate energy in the form of ATP through their outer membranes, just as bacteria and archaea do. The cells then use this energy currency to perform their activities. Suddenly growth is possible, because, as the cell grows larger, the mitochondria multiply, providing more energy to the cell. With an internal source of energy, eukaryotic cells can grow as large and complex as they want (well, pretty much) and, most importantly, evolve into multicellular organisms.

Interestingly, mitochondria look and function very much like bacteria, even having their own little free-floating ring of DNA. That should not be surprising, though. Just as chloroplasts in plants and algae were once free-living cyanobacteria, mitochondria are the descendants of a once free-living bacterium. It is surmised that, one day about 1.8 billion years ago, a single bacterium and one archaeon were living in a symbiotic relationship, each providing a survival advantage to the other. Such mutually beneficial relationships are common amongst microorganisms. What is not common is what happened next. At some point, the bacterium actually entered the archaeon. One prokaryote being engulfed by another is, to say the least, a rare and highly improbable event. This is particularly so since prokaryotes are not capable of phagocytosis, that is, the process by which a cell ingests or engulfs (eats) another cell or particle. How the bacterium got inside the archaeon remains a mystery. It did, however, happen, and somehow both survived. The symbiotic relationship continued, now with the bacterium providing the archaeon with an internal source of energy and the archaeon providing the bacterium with protection, a stable environment and nutrients. And, as is so often said, the rest is history.

So, with the pieces of the puzzle now all in place, let's step back and take a look at the completed picture. At the very beginning of life on this planet, say around 3.5 billion years ago, the bacteria and the archaea diverged from a common ancestor. This was at a point so early in the emergence of life that, among other things, each had to separately coble together its own type of cell wall and outer membrane and devise its own metabolic pathways. Then there they sat for a couple of billion years, developing all sorts of interesting biochemistry but unable to grow in size or complexity. About 1.8 billion years ago, a bacterium and an archaeon living in a symbiotic relationship somehow merged together, forming a single organism where there were once two. Now, every living thing on this planet that is not bacteria or archaea, is the product of this merger. There were, then, no intermediaries. Eukarya are composite creatures; archaea with internalized bacteria.

For present purposes, the significance of this story is that, as with the development of oxygenic photosynthesis, it happened only once. Without this impossibly improbable union, the planet would have remained a microbial world. But it did happen, and as a result we are here to contemplate its meaning. We have another winner.

Assessment. The three most important events in evolution were the appearance of oxygenic photosynthesis and of the complex cell, along with the emergence of life itself. We know that the first two evolved only once, and we have evidence of the third having happened only once. One-time events introduce an element of chance into the equation, particularly when they take billions of years to occur and do so for no discernable reason.

Both photosynthesis and the complex cell are absolute prerequisites to the evolution of intelligent life, yet both were the product of freakish accidents. Accidents, however, can happen and did happen. Can they be expected to happen on another planet? Who can say?

3.   Life Unfolding

The Boring Billion. 1.8 billion years ago the anoxic, microbe-dominated water-world that had persisted for two billion years was, at last, beginning to change.

A hundred million years or so after the Earth's encounter with Theia (4.5 billion years ago), the planet had cooled sufficiently for a thin, hard crust of basalt to form. Below, superheated magma rose to the surface while the cooler, near-surface magma sunk into the depths, forming vast convection cells thousands of miles across and hundreds of miles deep. Initially the crust and upper mantle (referred to together as the lithosphere) formed a single, unbroken plate, much like the surface of Venus today. Eventually, however, the pressure from below shattered the single plate into a number of separate plates.

By about 3.2 billion years ago, the interior of the earth had cooled considerably, resulting in a more organized mantle, with dozens of smaller, more stable cells replacing the massive convection cells. Since basalt is less dense than magma, the plates floated on the magma ocean below, pushed along by the constant churning of the mantle's convection cells. This was the beginning of plate tectonics.

When two tectonic plates converge, one flows beneath the other in a process called subduction. The subducted plate then melts again. The magma formed by this melted basalt is cleansed of iron and other heavy elements and becomes enriched with silicon. Being less dense than the surrounding basaltic magma, this silicon-rich magma slowly ascends, eventually crystalizing to form granite. In this way, granite islands began forming along the subduction zones, riding atop the denser basaltic crust like a cork on water. Over time, the granite islands piled up to form the first continents. This process continues to the present. Today, land makes up about 29% of the Earth's surface. 1.8 billion years ago that percentage was much lower, perhaps one quarter to one third of what it is now. The world, though, was beginning to look more like it does today.

By this time the cyanobacteria had been practicing their unique brand of photosynthesis and pumping out oxygen for half a billion years. The highly reactive gas, however, had to react with everything it could possibly react with before it could begin to accumulate in the atmosphere and the oceans. Thus, by 1.8 billion years ago, oxygen comprised only one or two percent of the air (as opposed to 21% today). This was far too low to permit the development of complex, multicellular life, but it was a start. And with so much of the Earth now already oxidized, would not further accumulations of oxygen in the atmosphere be just around the corner?

It was at this same time that the serendipitous union of two symbiotic prokaryotes had created the first eukaryotic life, life that would be able to use the volatile oxygen to power its drive toward size and complexity.

And so, at 1.8 billion years ago, all of the elements were in place for dynamic, diverse and abundant life to flourish. Or so it would have seemed. But then something happened. Something that would have been wholly unforeseen and unsuspected. Something that completely changed the destination that the world seemed to have been travelling towards. What happened? Nothing happened. The world had entered an era of stability, immutability, stasis, and nothing much would happen for a billion years.

What could possibly have happened to cause nothing to happen and to continue not to happen for such a very long time?

For starters, there is plate tectonics. The granite continents atop their basaltic plates drift around the globe like so many unpiloted ships. As you might expect, some of the continents will eventually collide and become wedged together. Like a multicar pileup on an interstate highway, other continents might crash into the wreckage. Such a logjam of all or most of the continents is referred to as a supercontinent. The most recent supercontinent, Pangaea, formed about 310 million years ago and broke apart about 130 million years later. The next is expected in about 200 million years from now.

At about 1.8 billion years ago, the first true supercontinent, variously referred to as Columbia, or Nuna or Hudsonland, was formed by such collisions. Columbia stretched 8,000 miles north to south and 3,000 miles east to west, incorporating virtually all of the Earth's continental crust. This supercontinent began to break up about 1.4 billion years ago, but a new supercontinent, Rodinia, reassembled itself by about 1.2 billion years ago and remained intact for another 400 million years.

Thus, for almost the entirety of the billion-year period from 1.8 billion to 800 million years ago, there was a single island continent that straddled the equator and was surrounded by a vast, world encompassing ocean. It is curious that the supercontinent persisted for so long. More recent continent conglomerations have broken up after only a hundred million years or so, quite a long time but only a fraction of the duration of Columbia/Rodinia. The answer appears to be that the Earth was then much younger and the mantle much hotter. The great interior heat of the planet softened the subducting crust, lessening its ability to pull more crust down behind it. It was not until about 800 million years ago that the Earth had cooled sufficiently to allow for today's more efficient subduction. When it did, Rodinia was ripped apart.

Both Columbia and Rodinia appear to have consisted of narrow coastal regions separated from the interior by mountain ranges that inhibited the flow of moisture from the ocean to the interior. Moreover, the absence of sedimentary rocks from this period indicates that there were no inland seas or waterways. And so, except for the coastal perimeter, the multi-continent island would have been a huge, hot, dry desert, much like the outback of Australia today, only worse. The sequestration of so much of the Earth's land mass from the ocean, significantly limited the amount and variety of habitats available for life and the deposition into the seas of much needed minerals and nutrients from the weathering of the rocky surface.

During this billion-year period, the waters of the boundless ocean were able to mix freely, homogenizing water temperatures worldwide. Also, the location of the supercontinent at the equator rather than at the poles prevented glaciation. These conditions combined to keep the weather constant and free of extremes. There also appears to have been only limited geologic activity. As one expert said of the period, "The atmosphere, the oceans and the crust of the Earth were acting as a stable interlinked system." It is not for nothing that this period is often referred to as the "boring billion."

The boring billion may have been a time of tranquility, calm and stability, but stability is good only when times are good. When they are not, it is not, and this was certainly not a good time for the unfolding of life.

It had long been assumed that the great oxygenation event that began with the development of oxygenic photosynthesis about 2.4 billion years ago had in due course fully oxygenated the oceans. In 1998, however, the Danish geologist, Donald Canfield, questioned this assumption. In an article published in the journal *Nature*, he proposed that sulfur, not oxygen, was the dominant factor in the Earth's oceans during the period from 1.8 billion to 800 million years ago. As you will recall, iron readily dissolves in water, provided no oxygen is present. When oxygen is present, it combines with the iron to form rust, which falls out of solution. The early oceans were anoxic and saturated with iron. In Canfield's view, enough oxygen had been produced by 1.8 billion years ago to significantly diminish the amount of dissolved iron in the oceans but not nearly enough to have oxygenated any but the near surface depths. The scant oxygen that had accumulated in the atmosphere, though, was more than sufficient to oxidize the abundant pyrite (iron sulfide) in the continental rocks, which, through weathering and transport by rivers, introduced a significant amount of sulfate into the ocean. Since sulfur reacts with iron to from insoluble pyrite, the sulfates removed the remaining iron from the seas. Thus, the ocean became stratified, with a shallow layer (perhaps only 10 to 20 feet) of clear, lightly oxygenated water atop a deep layer enriched in sulfur but essentially free of oxygen and iron.

The deep anoxic, sulfidic Canfield Ocean (as the ocean of that time is now termed) was a bad place for life, other than for purple and green sulfur bacteria, for which it was ideal. These bacteria engage in anoxygenic photosynthesis, using the energy of sunlight to reduce sulfates into hydrogen sulfide, which is both poisonous to eukaryotic life and stinks of rotten eggs and raw sewerage.

There was another problem for the eukaryotes as well. Nitrogen is essential to life. Nitrogen gas (N_2) is abundant, comprising 79% of today's atmosphere, but life's biochemistry cannot utilize nitrogen gas. It can, however, use ammonia (NH_3). Life has evolved an enzyme called nitrogenase that converts nitrogen gas to ammonia. To do so, the enzyme utilizes a cluster of atoms of sulfur and a metal, either iron or molybdenum. Therein lies the problem. By this time, the iron had been removed from the ocean. Moreover, Molybdenum is soluble only in oxygen rich waters, but the ocean was, for the most part, anoxic. Thus, life could thrive only in coastal waters, where weathered molybdenum was washed into the sea.

So that was the world where the first eukaryotic life came into existence: a suffocating, lightly oxygenated atmosphere that stunk of raw sewerage; a scorching, dry island continent, barren and lifeless; and a limitless, anoxic ocean devoid of life, except in coastal waters which were split between a deep, poisonous, anoxic layer ruled by purple and green sulfur bacteria and a shallow, poorly oxygenated layer of clear water, where algae and cyanobacteria could still produce a bit oxygen and where a limited population of simple, single-cell eukaryotes could, if not evolve and develop, at least cling to life. And there it sat, unchanging and stagnant, day after day, week after week, month after month for a thousand million years.

Snowballs From Hell. 800 million years ago the supercontinent, Rodinia, was at last breaking apart. Modern plate tectonics was now underway, busting up the billion-year continental logjam and sending the individual continents off on their separate paths.

The breakup created numerous shallow seas and inland waterways, exposing to the weather millions of square miles of land that had previously been sequestered within the vast, parched interior of the supercontinent. Tropical rains washed minerals, including large amounts of iron, into the sea. There, the iron reacted with dissolved sulfurous compounds to form pyrite (iron sulfide), which, being insoluble, fell out of solution. With the depletion of sulfur from the seas, the billion-year reign of the purple and green sulfur bacteria as the Earth's dominant life form came to an end. The shallow, sun drenched seas and extensive new coastal waters were perfect habitats for algae. The abundant minerals caused massive algal blooms that pumped oxygen into the atmosphere.

With the shattering of Rodinia, the ending of the Canfield Ocean, the resumption of the oxygenation of the atmosphere and a profusion of simple life forms, all seemed set for the advance toward complex life. But then, once again, the unexpected happened.

Glaciers crush and scour the surface of the continents as they advance, leaving behind thick, irregular layers of tillites, composed of sand, gravel and rock fragments, along with scattered boulders and moraines. As a result, evidence of past glaciation is obvious to the trained eye. Geologists have long been aware of such glacial features in rocks dating between

750 million and 600 million years. In fact, almost all rocks of that age show evidence of glaciation. In the 1990s, a team of researchers found tillites of that age that had been laid down at sea level just a few degrees from the equator. That could only have meant that glaciers had once spread all the way down to the equator. Even at the height of the last ice age (actually the present ice age, since we are now only in an interglacial period that will end in a few thousand years), glaciers never advanced much farther south than 45 degrees north latitude, so evidence of glaciers at the equator was a shocking find.

What could have happened to have caused glaciation on such an enormous scale? Well, it seems that, after Rodinia, the continents remained in equatorial waters. The tropical rains did more than just wash minerals into the seas. They weathered the formerly dry, barren rocks. The weathering exposed silicate rock, which then soaked up carbon dioxide out of the air like a sponge, significantly diminishing the amount of CO_2 in the atmosphere. At the same time, the blooms of photosynthesizing algae were profligately taking in CO_2, further drawing down CO_2 levels. As the stores of this important greenhouse gas plummeted, the Earth grew colder. Eventually ice began to form at the poles. The dark oceans absorb sunlight, bringing warmth to the world. Ice, on the other hand, reflects sunlight back into space. This process, referred to by astronomers as the albedo effect, has a significant impact on the climate of a planet. Thus, the presence of ice caused the Earth to cool, which caused more ice to form, which caused further cooling. Since the continents were then near the equator, the weathering of rocks continued, further diminishing CO_2 levels, even as the ice was growing. Eventually a tipping point was reached, and, by 715 million years ago, the ice had spread to the equator, encasing the Earth in a mile-thick layer of ice. Such snowball Earth conditions continued for some 50 million years.

During this time, little light would have been able to penetrate the icy sheath. That would have brought an end to photosynthesis and to eukaryotic life, leaving only the populations of prokaryotes then living around the vents in the still anoxic ocean depths. Since that did not happen, it is probable that there remained numerous, perhaps thousands, of scattered ice-free ponds surrounding volcanoes, geysers and hot springs. Though almost all life on the planet succumbed to the freezing conditions, limited populations of prokaryotes, single celled eukaryotes and algae managed to hold on in these isolated refuges. Isolation is one of the great drivers of evolution. So, in a case of "what doesn't kill me makes me stronger," the snowball Earth would have stimulated extensive diversity amongst the survivors, which would be of crucial importance to the long-term viability of complex life.

The snowball Earth theory appeared to be consistent with the fossil and geologic records, but there was one glaring problem. Once the snowball was formed, there seemed to be no way out. The early Sun was only about 75% as luminous as it is today. Since that time, its luminosity has increased by about seven percent per billion years. At that rate, it would have taken several billion years for the Sun to have grown hot enough to melt the ice. Clearly, an expanding Sun was not the answer. In an "ah-hah" moment, it occurred to one researcher that the conditions on the surface of the planet would have had no effect whatsoever on what was transpiring in the thousands of miles of molten rock and metals in the interior. Volcanoes would continue to grow and erupt, pouring tons of CO_2 into the atmosphere, whether or not the surface was covered by a veneer of ice. Over time, CO_2 concentrations would have increased to hundreds of times modern levels, blanketing the planet in the warmth-preserving greenhouse gas. Moreover, since the ice prevented the weathering of rocks and since carbon dioxide consuming algae would have been so significantly diminished, there was nothing to prevent CO_2 from continuing to accumulate. Over time, the atmosphere warmed, eventually growing so warm that the ice began to melt. Reversing the freezing cycle, the melting of the ice reduced the area that reflected sunlight and increased the area that absorbed it. Thus, the melting of the ice caused the planet to grow warmer, which caused more ice to melt, which caused the Earth to warm and so on and so on, you get the picture.

In this way, the snowball, with an average temperature of -50 degrees F, turned into a hothouse, averaging 50 degrees C (122 F). And it all took no more than a thousand years to do so.

It didn't stop there, though. The high temperatures caused more evaporation, which caused more rain, which caused more weathering, which caused the depletion of atmospheric CO_2, which caused temperatures to drop, which caused ice to form, which reflected sunlight, which caused the temperatures to drop further, which caused more ice to form, and soon the Earth was right back in the snowball.

If you guessed that another hothouse followed, you'd be right. There were two major snowball events and a couple of minor ones, but, by 635 million years ago, the hundred-million-year roller coaster ride had come to an end and more stable conditions prevailed. Why it did is subject to debate. Perhaps the continents moved more northerly. If so, the formation of ice over land, though still reflecting sunlight, would stop the weathering of rock and the resulting depletion of atmospheric CO_2. There are other ideas, but that is a little beyond the scope of our present inquiry.

For our purposes, the snowball cycle is important because of the changes it made to the planet. Advancing glaciers ground rocks down to rubble. The melting of the glaciers and heavy rains swept millions of tons of iron, nitrates and, most importantly, phosphates into the seas. Today these minerals are the components of fertilizers. 700 million years ago they served the same purpose. This fertilizer caused successive waves of algae blooms that pumped oxygen into the atmosphere at a prodigious rate. Since the world was well oxidized by that time and there were no animals to devour the algae, oxygen accumulated to unprecedented concentrations. By the end of the snowball cycles, the oxygen level had risen substantially, though not nearly to today's levels.

And so, after nearly 4 billion years, was life finally ready to take a great leap forward?

Animal Kingdom. The eukaryotes consist of single-celled protists, algae, fungi, plants and animals. Of these, only animals would seem to have the wherewithal to develop intelligence. So, in this section, we will examine the rise of animals. But not all animals.

The animal kingdom is divided into two major branches. The first and oldest branch would be the sponges (phylum porifera). Sponges are sessile (attached and immovable) animals that grow into a vase-like shape with an opening at the top. The inner surface is lined with collar cells having hair-like flagella that beat in unison to draw in water and food particles and to expel waste. They have no organs. There are between 5,000 and 10,000 species of porifera.

All other animals are grouped into two further branches, the cnidaria and the bilateria. The cnidaria, which include jellyfish, corals, sea pens and sea anemones, are radially symmetrical. They are shaped like a sack with tentacles surrounding a mouth and are composed of two cell layers sandwiching a gelatinous material. They are the most primitive animals having distinct tissue (including muscle and nerve), though they also have no organs. While sponges simply filter organic particles out of the water, Cnidarians hunt with tentacles armed with stinging cells. There are about 9,000 species of cnidarians.

All of the remaining ten million or so species of animals are bilaterians. Bilaterians are bilaterally symmetrical, having a top and a bottom, a front and a rear and two roughly identical sides. They have not two, but three cell layers. Like the cnidarians they form tissue, but unlike them, they have organs as well.

Cnidarians may have invented predation, but bilaterians perfected it. To actively hunt, an organism must move, which requires muscles. To move effectively, muscle contractions must be coordinated, which requires a nervous system. To locate prey, the organism must have sensory organs. To coordinate movement and to interpret sensory perceptions, the organism must have a brain. That is what we are looking for. We will find them in fossil-bearing rock.

Most rocks at the surface are sedimentary, having been formed by older rocks that have been broken apart by wind or water. Gravel, sand and mud settle to the bottom of rivers, lakes and oceans. As the material is further overlain by more sediments, the building weight and pressure causes the bottom sediments to form rock. When different kinds of sediments are laid down over existing sedimentary rocks, layers are formed. Sedimentary layers that are identifiably different from each other are called strata. Over time, dozens of strata may stack one on top of the other. In 1669 the Danish physician, Nicholas Steno, made the rather obvious supposition that, in any sequence of sedimentary rocks, any one layer is older than the one above it and younger than the one below it. This observation is now known as the law of superposition.

The sediments forming the strata often include the remains of the creatures that had lived in the environment. The soft tissue rapidly decays, but the hard parts fossilize, leaving a record of life as it existed at the time the sedimentary strata were laid down. Strata in different areas around the world that contain the same or similar fossils were likely formed at about the same time. The identification of strata by the fossils they contained was initiated by William Smith and Georges Cuvier in the early nineteenth century. Their work enabled geologists to correlate strata in different areas and even different continents. Detailed studies of strata and fossils between 1820 and 1850 produced the sequence of geologic periods still in use today.

The bottommost, and, therefore, most ancient, of the fossil bearing strata was in 1835 given the name Cambrian by the pioneer geologist, Adam Sedgwick, from the region of Cambria in Wales where he was studying the stratum. The Cambrian stratum was loaded with fossils of diverse forms of animals, largely arthropods, such as the iconic trilobites, but others as well, including mollusks, brachiopods and echinoderms. Many fossils were of large, complex animals. The trilobites, for instance, were segmented, with well-developed limbs and eyes. Beneath the Cambrian, however, the strata were devoid of any visible fossils.

The sudden appearance of large, complex life in the fossil record without precursors was a matter of deep concern for Charles Darwin. His theory of descent with modification and natural selection assumed life evolved gradually over extended periods of time from simple to more complex forms. The seeming explosion of life *ex nihilo* in the Cambrian period was a serious obstacle to his theories and supported creationist beliefs. In 1859, in his *Origin of Species*, Darwin wrote, "To the question of why we do not find rich fossiliferous deposits belonging to these ... periods prior to the Cambrian System, I can give no satisfactory answer." He went to his grave, however, confidant that the fossil record would sooner or later reveal the missing simpler life forms from which the Cambrian fauna were derived. And he was right... sort of.

In 1946, a state geologist working in the desolate Ediacaran Hills of South Australia happened upon what appeared to him to be casts or impressions of jellyfish in scattered sandstone slabs that he knew to be very ancient, perhaps older than the Cambrian rocks. The creatures came in a great variety of forms, though all differed from any known organism, living or fossil. Known today as Ediacaran biota, they have since been found in some thirty localities in six continents and classified into seventy different species. We now know they exploded onto the scene about 575 million years ago, that is, about 60 million years after the last snowball event and 35 million years before the beginning of the Cambrian Period (541 to 488 million years ago (MYA)). Then, just as suddenly, they disappeared from the fossil record right at about the beginning of the Cambrian period.

Were these ediacarans Darwin's missing antecedents to the Cambrian fauna? That was difficult to say since their shapes and morphology resembled no animal since. Some came in disks that apparently laid upon the microbial mats that then carpeted the ocean floor. Others, known as rangeomorphs, resembled underwater leaves or ferns that also rooted themselves into the mats. Cloudina consisted of calcareous cones, stacked one inside the other. The dickinsonia were oval in shape with segments alternating from a midline in what is termed a glidepath symmetry, rather than bilateral symmetry, which has never been found in any other animal, except other ediacarans. Not to be deterred, scientists, looking for any similarities, began shoehorning the ediacarans into various modern animal taxa. The discs were seen as impressions of jellyfish, the rangeomorphs the ancestor of sea pens or sea anemones, the dickinsonia early worms, kimberella mollusks. Others ediacarans were acclaimed as the first sea urchin or the first crab, and on and on.

But were they? Significant differences between the ediacarans and all other animals became increasingly hard to ignore. For starters, none had a mouth, nor for that matter jaws or a gut, or even an anus. None had any evidence of organs or any apparent means of locomotion. Dickinsonians could range in size from as small as a coin to as large as a turkey platter, but they were never more than a few millimeters in width. How the edaicarans subsisted, we have no clue. Without a mouth and jaws or means of locomotion, they certainly did not graze or hunt. Some argue they may have been filter feeders, like today's sponges, or early forms of cnidarians that captured their prey, though they had no flagella or tentacles or stinging cells or even a mouth with which to consume them. Some suggest they photosynthesized or engaged in symbiotic relationships with algae, as some corals do today. This, however, is unlikely since many ediacarans lived on the floor of the deep ocean, far beyond the reach of sunlight. Others proposed that they absorbed nutrients from the sea water by osmosis or from the mats and sediment upon which they lay, but without proof or suggestion as to how that might have worked.

Most scientists now agree that they were likely some sort of early animal, or colony of animals, with probably no surviving descendants. One expert even claims they were completely unrelated to animals and formed their own kingdom of creatures that no longer exists today; just an early failed experiment in evolution.

I think the attempt to connect the ediacarans with the modern animal types that arose in the Cambrian by looking for structural similarities missed the point. Cambrian jellyfish, for instance, developed tentacles and stinging cells to capture the zooplankton on which they lived, and sponges developed flagella to create currents to draw in bacterial particles that had been stirred up by burrowing worms in the sediment. Since there were no zooplankton or worms in the Ediacaran period, it would have been pointless for the biota to develop those features. The ediacarans were, apparently, well adapted to the environment as it was at the time they lived. Indeed, they were the dominant life form on the planet for nearly 35 million years, hardly a failed experiment. But, as we will see, the world changed, spelling their demise.

Near the beginning of the Cambrian period, the fossil record begins showing tracks in the sediment of tiny, burrowing, wormlike creatures. It is probable that these newly appearing animals would have quickly disrupted or devoured the microbial mats upon which the ediacarans lived. Furthermore, shells of all sorts suddenly become abundant. Since the purpose of a shell is protection, it is clear that predation was then underway. The sessile, unprotected ediacarans would have been defenseless against the onslaught of the hunters. As a result, they quickly disappeared.

That then brings us back to where we started, a sudden profusion of animal life in the Cambrian, without evidence of direct ancestors in the fossil record. Yet, absent special creation by the hand of God, there must have been simpler animals at an earlier time.

Though the fossil record is not much help in this regard, there is another way to determine when the earliest animals evolved. DNA and protein sequences evolve at a rate that is relatively constant over time and amongst different organisms. Thus, the genetic differences between any two species is proportional to the time since the species shared a common ancestor. By measuring the genetic differences between genes in two species whose time of divergence from a common ancestor is known from the fossil record, the length of time for any single genetic difference to occur can be estimated. When this rate of change is multiplied by the number of genetic differences between two species whose time of divergence from a common ancestor is not known from the fossil record, the time of divergence can be estimated.

Since its introduction in the 1960s, this "molecular clock" method of estimating dates of divergence has grown more sophisticated and accurate. Recent studies estimate that sponges and other animals diverged from some stem ancestor of all animals about 700 million years ago and that the proto bilaterals and cnidarians diverged about 650 million years ago. These estimates may change as methods improve, but it is clear that the early branches of animals existed well more than 100 million years before animals begin to appear in the fossil record.

So again, Darwin was right... kind of. Simple animals apparently existed long before the appearance of fossils of complex animals like trilobites in the Cambrian. But if so, where were they?

The answer, it appears, is that these early animals were minute, gossamer and without hard parts. As a result, they were not likely to be preserved in the fossil record. Nematodes (round worms), for example, are today the most numerous multicellular animal on the planet. A square yard of soil may contain as many as a million of these microscopic worms, and they are found in almost every environment, whether terrestrial, freshwater or saltwater. Nematodes have been around since the Cambrian period, yet they have left almost no trace in the fossil record. Likewise, the earliest animals could be hypothesized, but it seemed unlikely that their existence would ever be confirmed by the fossil record.

Then in a 2005 paper in *Scientific American*, a University of Southern California paleontologist announced the finding of a tiny (as long as a human hair is wide) fossil of a bilateral animal in the Doushantuo formation in southern China. The formation was dated to 600 million years ago, and thus the fossil, christened vernnanimalcula, was by far the oldest bilateral ever found. A missing link that was missing no longer. On further review by other researchers, however, it was determined that the fossil was neither a bilateral, nor an animal nor probably even a fossil. In his paper, the discoverer had boasted that his group had gone to China with the intent of finding microscopic fossils of the earliest bilateral animals. And, by golly, that is exactly what they found, or thought they had found. Once again, people see what they want to see and believe what they want to believe, scientists included. And so it goes.

Despite such unfortunate errors and the unlikelihood of ever finding supporting fossil evidence, it is highly probable that the sponges, cnidaria and bilaterials all spent many millions of years as simple, microscopic creatures before growing and differentiating in the Cambrian. The question, then, is why? The answer... oxygen. It is well established that the abundance and diversity of animal species in modern marine basins declines precipitously as oxygen levels drop. Nematodes, on the other hand, can get along just fine in the nearly anoxic conditions of the sea floor sediment because of their simplicity and tiny size.

As previously discussed, oxygen levels climbed significantly at the end of the last major snowball event about 635 million years ago, levels high enough, apparently, for the evolution of simple, stem animals, but only just so. There is also evidence of another rise in oxygen about 575 million years ago, possibly as a result of another, lesser snowball event. The increased oxygen levels likely allowed the multicellular

Ediacaran biota to emerge, though levels remained substantially lower than they are today. Then about 550 million years ago, and for reasons not clearly understood, oxygen levels rose once again, this time to about 13% (compared to 21% today), sufficient to enable animals to grow and diversify. And so they did.

Darwin, then, was right in his belief that Cambrian animals were preceded by simpler forms but wrong that this would be borne out by the fossil record.

The sudden appearance of a profusion of animal fossils in the Cambrian formation seemingly from out of the blue is generally referred to as the "Cambrian explosion." Since the time the term was coined, however, scientists have argued whether the appearance of animal life was really all that sudden. After all, the Cambrian period began 541 MYA, but the first trilobite fossils do not appear until 521 MYA, and animal groups continued to diversify until the end of the period 488 MYA. To many, 50 million years is not terribly sudden. However, it was during the Cambrian period that all thirty-two or so animal phyla came into existence (phyla being the largest taxonomic rank below kingdom, followed by class, order, family, genus and species). Since Cambrian times, not a single animal phylum has been lost or new one added. When stacked up against the largely uneventful 3000 million years of life that preceded it, I'd say 50 million years looks pretty sudden.

So how is it that all of the animal phyla could have arisen in such a comparatively short period of time? The answer, I think, is that's how evolution works.

Darwin believed that species evolved gradually over long periods of time, slowly and incrementally morphing into new and different forms. That's not, however, what the fossil record shows. Instead, fossils that are abundant in one stratum are often absent in the very next, with entirely new fossils abruptly taking their place. Then, in the next stratum, the new fossils are replaced by yet others. And on it goes throughout successive strata.

This is because natural selection drives species to adapt to their environment. To state the obvious, individuals better suited to their environment are more likely to survive and reproduce than those that are less suited. Over time, a species reaches an optimum level of adaptation. By that time, however, the continuous culling of less adapted individuals has caused the genetic make-up of the species to become very homogenous. Thereafter, natural selection ceases to act as a tool of adaptation and becomes a ruthless enforcer of conformity. That is because virtually all genetic mutations (or, more often, mutations in gene expression, i.e., DNA instructions for making proteins or other functional molecules) cause the individual to become less adapted and thus less able to compete for limited resources. Thus, when competition is heavy, individuals with genetic mutations do not fare very well and soon perish, taking their mutations with them. Consequently, species may remain fixed for thousands or even millions of years.

Sooner or later, however, the environment changes. Species once well adapted to the previous environment will suddenly find themselves poorly adapted to the new environment, causing individuals and even entire species to perish. When that occurs, competition for resources weakens. Now, an organism with a genetic mutation that makes it less fit (but doesn't kill it) can survive. Offspring of the mutant individuals may have further mutations of their own, and they too can survive. Before long, a great variety of genetic mutations collect in the population. As populations begin to increase, however, so does competition. Natural selection then begins favoring those that are better adapted to the new environment, reducing genetic diversity. When the ecosystem is once again fully populated, new species become fixed and another period of stasis ensues. This pattern of long periods of stasis followed by rapid bursts of evolutionary change followed again by stasis was given the name "punctuated equilibrium" by paleontologists Niles Eldredge and Stephen Gould.

Evolutionary biologists refer to an environment where populations have been diminished by environmental change as a "permissive ecology." It is permissive because competition is weak, permitting genetic diversity and novelty to develop. Permissive ecologies are a regular feature over geologic time, but never was an environment more permissive than it was at the beginning of the Cambrian. It was a time when the world was vacant and anything was possible. Animals quickly devised all manner of body plans to exploit available resources. The successful general body plan of a phylum engendered a host of species, all being just variations on the theme. Once the world was fully populated, however, such fundamental innovations came to an end, leaving us with the animal phyla as we have today.

Life Marches On. Another half a billion years would pass before the more recent times that will be the subject of Part III of this assessment. Although this is not intended to be a study of paleontology, a brief overview of the development and spread of life during that time would, I think, lend context to future discussions.

As noted above, the modern geologic time scale was devised by geologists and paleontologists between 1820 and 1850. They divided fossil bearing strata into three eras, the Paleozoic (old life), the Mesozoic (middle life) and the Cenozoic (recent life), each of which was further subdivided into periods, the first being the Cambrian Period. More ancient strata contained no visible fossils so were dismissed as azoic (without life) or simply referred to as the Precambrian. In the twentieth century, with the recognition that life had been around for possibly as long as four billion years and that much had transpired in the Precambrian world, the geologic history of the world was divided into four eons, those being the Hadean (4560 to 4000 MYA), starting from Earth's beginning and continuing until about the end of the late heavy bombardment; the Archean (4000 to 2500 MYA), continuing until the first oxidation event; the Proterozoic (2500 to 541 MYA), continuing through the boring billion, the snowball events and the Ediacaran down to the Cambrian era; and Phanerozoic (visible life), continuing to present times.

A quick note on the dates of these geologic time periods. Nineteenth century geologists knew the relative dates of the strata they were studying, that is to say they knew whether a particular stratum was older or younger than another. The actual age of the rocks, however, they had no clue. In the early twentieth century, a method of dating rocks by measuring known decay rates of radioactive isotopes was developed. The method is known as "radiometric dating," and this is how it works.

Atoms of each element may vary slightly in the number of neutrons in their nuclei. These variants are called isotopes. Some isotopes are radioactive, meaning they are unstable because their nuclei are too large. Over time, unstable isotopes will lose either neutrons or protons and change or "decay" into a stable isotope. The rates of decay for many radioactive isotopes have been measured and are known. Fortunately, each kind of radioactive isotope decays at its own fixed rate, some having shorter periods, such as carbon-14 to nitrogen (with a "half-life," or the time it takes for half the radioactive carbon-14 isotopes in a given sample to decay into stable nitrogen, of 5,730 years), others longer, such as uranium-235 to lead-207 (with a half-life of 704 million years), and some longer yet, such as potassium-40 to argon-40 (with a half-life of 1.3 billion years).

When volcanic (igneous) rocks form, they trap radioactive elements that will decay over time into stable atoms. Thus, igneous rocks can be accurately dated by measuring the ratio of the stable and remaining unstable isotopes they contain and applying the rates of decay. Sedimentary rocks (where the fossils are) are dated by dating igneous rocks or volcanic ash just above and below the sedimentary strata, giving an approximate date of the strata.

And now to the Phanerozoic Eon, the time of visible, multicellular life forms. For most of us, this is the time of greatest interest and importance, since it is the time of the rise and development of what we consider real life, plants, fungi, animals and, of course, ourselves. We should bear in mind, however, that the Earth had been teeming with life for some three billion years prior to the beginning of the eon 541 million years ago. Indeed, the prokaryotes (bacteria and archaea) still overwhelm us numerically, filling to capacity every nook and cranny of this planet. A liter of seawater, for instance, will contain about a billion bacteria (not to mention 10 billion viruses). What's more, our own bodies contain about 30 trillion cells but 39 trillion bacteria. This is and has always been a microbial world. If the bacteria and archaea were to disappear tomorrow, life on this planet would grind quickly to a halt. If, on the other hand, we multicellular life forms were to all perish, the microbes would hardly notice our absence. That, though, is another story, so let us proceed with our whirlwind tour of the Phanerozoic.

The Paleozoic Era covers a span of about 300 million years, divided into six Periods, the first of which was the Cambrian Period (541 to 488 MYA), that we previously discussed. The Cambrian was followed by the Ordovician

Period (488 to 444 MYA). That period witnessed the assembly of the supercontinent, Gondwana, which was formed from the present-day continents of Africa, South America, Antarctica and Australia, with little land north of the equator. Carbon dioxide levels were high, many times the levels of today, and the climate was warm. As a result, continents were flooded to an unprecedented level. There was a significant increase in the number of species, and the seas were dominated by invertebrates, such as bryozoans and brachiopods.

The end of the period was marked by a mass extinction, the second greatest of all time. Nearly ninety percent of all species were lost. The event was likely the result of extensive glaciation as a result of Gondwana having drifted over the south pole.

The Ordovician was followed by the Silurian Period (444 to 416 MYA). The warm, stable climate of the Silurian melted the glaciers, causing sea levels to rise again. Much of present-day North America was covered by a shallow sea, creating new habitats for marine life. There is evidence of extensive reef building and the beginnings of fresh water and estuarial ecosystems.

For our purposes, the most momentous event of the Silurian was life's first efforts to colonize dry land. Cooksonia was the first plant with stems and vascular tissue for water transportation, though without leaves, appeared. Also, the first air breathing animals developed, all arthropods, including millipedes, centipedes and arachnids.

Next came the Devonian Period (416 to 358 MYA), sometimes referred to as the age of fishes for the rapid proliferation of primitive fishes. Placoderms (armored fishes) became the dominant marine predators, some growing up to 30 feet in length. Also, new types of fishes, the ray-finned fish and the lobe-finned fish, having true bones, teeth, swim bladders and gills, made their first appearances. On land, horsetails, ferns and lycophytes grew to large sizes and formed the first forests. By the end of the period, the proliferation of plants had increased oxygen levels and depleted carbon dioxide. The first insect, the flightless rhyniella, evolved from marine crustaceans. Early tetrapods, such as tiktaalik, the probable link between lobe-finned fishes and amphibians, took their first tentative steps on land.

The close of the Devonian witnessed the second of the big five mass extinction events on this planet. Over a prolonged period, seventy to eighty percent of all animal species died off, probably as a result of global cooling resulting from depletion of carbon dioxide caused by the rapid spread of the first forests.

Marine and freshwater fishes continued to diversify in the Carboniferous Period (358 to 298 MYA). The first true bony fishes and the first sharks, which would dominate the seas, evolved during this period. On land, great forests covered the Earth, and giant swamps filled low-lying areas. The vast amount of vegetation caused atmospheric oxygen to reach the highest levels ever on this planet, perhaps as high as thirty-five percent (compared to today's twenty-one percent). Insects developed flight, with mayflies and dragonflies common. The increase in oxygen allowed some land arthropods to attain stupendous size, with millipedes as long as five feet and some dragonflies with wingspans of 30 inches. In the swampy environment the first amphibians flourished, some growing up to twenty feet in length. Late in the period, the first amniotes emerged, a group of limbed vertebrates that includes all living reptiles, birds and mammals and their extinct relatives.

The name Carboniferous, Latin for "coal-bearing," was given to the period for good reason, since ninety percent of the world's coal reserves date from around this period. That was the result of vegetation being buried beneath the swamps on an enormous scale. In fact, the rate of carbon burial was 600 times faster than at any other period Over time, heat and pressure transformed the buried organic material into coal.

Why was so much organic material buried during this Period but not afterward? Well, today, when a tree falls in the forest, new tenants rapidly move in. Bees, wasps, carpenter ants and wood-boring beetles carve tunnels into the dead wood to nest and lay their eggs, bringing with them wood digesting fungi; bark beetles and wood lice loosen the bark, allowing fungi to spread under the surface; and wood roaches and termites, with the aid of symbiotic microorganisms in their gut, feast on the wood. Soon, the tree has been recycled back into the woodland environment. In the Carboniferous Period, however, these insects did not yet exist. Indeed, the ability to digest the tough, fibrous lignin and cellulose had yet to evolve. So, the fallen tree might just lay there like a rock for decades, until its eventual burial. Moreover, since the buried organic material was not oxidized to carbon dioxide, the oxygen remained in the atmosphere, further contributing to the high oxygen levels at the time.

At the beginning of the Permian Period (298 to 252 MYA), glaciation was widespread. The assembly of the small continents into the supercontinent, Pangea, however, brought extensive hot, arid conditions and squeezed out shallow seas and estuaries that are the most productive part of the marine environments. Once again, there was only a single continent, surrounded by a world-wide sea.

Plants diversified and spread extensively, while insects evolved rapidly into the new habitats. Major reptile lineages evolved, including protosaurs, which would give rise to dinosaurs, crocodiles and birds; eosuchians, ancestral to snakes and lizards; and early anapsids, ancestors to turtles.

The other major group of amniotes, the synapsids, proliferated and diversified. The earliest and most primitive synapsids were the pelycosaurs, such as the lizard like, sail backed Dimetrodon. Midway through the Period, the pelycosaurs were succeeded by the therapsids, which would become the dominant large land animals and eventually give rise to mammals.

It was during the Permian that the first large herbivores appeared. Vegetarianism as a lifestyle posed significant problems for vertebrates. Plant cells are composed mostly of cellulose, which can be broken down only with certain enzymes that vertebrates do not possess. To counter this problem, herbivores use microbes, which they hold in fermenting chambers in their gut. How this first occurred is unclear. The evolution of herbivory, however it happened, was revolutionary. For the first time, vertebrates could directly access the vast resources provided by plants. The herbivores themselves then became the major food source for large terrestrial predators.

The end of the Permian witnessed another mass extinction event, which was the worst of them all. Over 95 percent of marine and 70 percent of terrestrial species disappeared. Near the end of the Period, a great plume of magma rose up through the mantle and melted the crust, turning the land mass that is today China into a cauldron of lava. Vast quantities of noxious gases were released into the atmosphere, acidifying the oceans and shredding the ozone layer. Five million years later, with life still reeling, a second, even greater magma plume broke the surface in an area of Russia known as the Siberian Traps, which cover some 500,000 square miles. Continuous eruptions over perhaps 60,000 years poured out enough lava to cover the continental United States to a depth of a mile, releasing huge amounts of ash and gas, which caused atmospheric levels of carbon dioxide, sulfur dioxide and methane to soar. Moreover, great quantities of lava intruded coal seams below ground, resulting in the release of even more greenhouse gases. Global temperatures climbed, reducing oxygen levels in the oceans by as much as 80 percent. Great quantities of carbon dioxide dissolved into the oceans, causing a sharp drop in the pH and making shell formation in the acidic waters difficult if not impossible.

By the end of the Great Dying, as this time is called, the Earth was nearly a dead planet. Most of the land was now desert, with vegetation sparse. The oceans were moribund, the reefs gone, the sea floor carpeted with a layer of stinking slime. Complex life managed to hang on, but just barely. Evolution of life on this planet was set back tens of millions of years, but it easily could have been hundreds of millions, illustrating the precarious nature of the entire endeavor.

The Paleozoic Era was over, and the Mesozoic Era had begun, the first period of which was the Triassic (252 to 199 MYA). Pangea remained the sole landmass, its vast interior remaining hot and dry. The depleted seas only slowly recovered. Early on a group of reptiles, the order icthyosauria, returned to the ocean, and, by the mid-Triassic, these streamlined, air breathing predators, some measuring fifty feet or more, dominated the seas.

On land, the hot, dry climate of the interior of Pangea limited advances by plants and insects. Probably due to its proclivity for burrowing, one animal species, Lystrosaurus, made it through the extinction event relatively intact. Built like a pig, but with a powerful head wider than it was long, massive jaw muscles and a sharp, horny beak, Lystrosaurus roamed the badlands of Pangea in vast herds, going anywhere and eating anything. With its adaptable, omnivorous lifestyle, this odd looking, porcine bulldozer would be the dominant, and nearly the only, terrestrial animal for millions of years, comprising perhaps ninety-five percent of all land vertebrates.

The reptilian archosaurs survived the Permian extinction and proliferated. By the mid-Triassic, one lineage evolved into the first dinosaurs and another into the first pterosaurs, probably then gliding rather than flying. Therapsids, known as mammal-like reptiles (though not reptiles at all), were dominant early in the Triassic but later nearly went extinct in the wake of the archosaurs. From the therapsids, however, evolved the first mammals. These earliest mammals were not terribly impressive, being only a few inches in length, mostly herbivores and insectivores and, probably arboreal and nocturnal as well. They would not be able to compete with the dinosaurs, which would keep them on the sideline for many millions of years. The period ended in another, though less devastating, mass extinction

The age of the dinosaurs begins in earnest with the Jurassic Period (199 to 145 MYA). As the period began, Pangea was at last breaking up. The continent's rifting apart and generally warm global temperatures promoted a great diversity of new life. The long-necked, herbivorous, quadrupedal sauropods proliferated. Some grew to over 100 feet in length and weighed over 100 tons. They were a highly successful, long surviving group that spread across every continent, except Antarctica, despite having a brain the size of a tennis ball. The apex predator of the period was allosaurus, which reached lengths of 30-40 feet and heights of 17 feet. Mammals remained tiny and shrew-like.

In the seas, cartilaginous and bony fishes were abundant and marine reptiles common. Plesiosaurs (aquatic reptiles) became the dominant predator, some growing as long as 50 feet. The pterosaurs dominated the skies. They were neither bird nor dinosaur but flying reptiles. They too could grow to enormous size, the largest, the quetzalcoatlus, having a wingspan of as much as 36 feet. Toward the end of the period, a number of species died out, probably as a result of climate change, but there was no mass extinction.

The final period of the Mesozoic was the Cretaceous (145 to 65 MYA). At the beginning of the period, Africa and South America were still joined, as were North America and Eurasia. By its end, the continents were moving toward their present configuration, and the globe would have been a much more familiar place. Global temperatures were much warmer than today, with tropical and subtropical conditions extended to 45 degrees N latitude and moderate temperatures even at the poles.

The Cretaceous saw the development and radiation of flowering plants. Birds, having evolved from two-legged theropod dinosaurs, took to the air, joining the pterosaurs. Sharks and rays, along with giant marine reptiles, such as plesiosaurs, mosasaurs and the fishlike ichthyosaurs, were common in the seas. Terrestrial life continued to be dominated by dinosaurs, which included large herds of the duck-billed dinosaurs known as hadrosaurs, elephant-sized triceratops and, of course, the fearsome tyrannosaurus rex, 40-50 feet in length and standing as tall as 20 feet at the hip (its highest point, since it did not walk upright). Many types of sauropods had died out, but the titanosaurs, which emerged in the second half of the period, were the largest land animals that have ever lived. Placental, monotreme and marsupial mammals evolved, but none ever exceeded the size of a rabbit.

Everyone knows that the Cretaceous Period came to an end 65 million years ago when an asteroid seven mile in diameter slammed into the Earth. The asteroid ejected vast amounts of molten rock and dust into the atmosphere, triggering world-wide wildfires and enshrouding the Earth in darkness for many months, shutting down photosynthesis and causing temperatures to plummet. In the late 1970s, geophysicists discovered the 110 mile in diameter Chicxulub crater off the Yucatan Peninsula in Mexico, which dates precisely to this time. Then in the 1980s, scientists found shocked quartz, tektites (tiny glass globes associated with meteorite impacts) and iridium (an element found only in extraterrestrial rocks) in strata around the world dating from this time, seemingly clinching the impact hypothesis. Yet even this iconic theory has been challenged, with some scientists blaming instead a huge outpouring of lava and associated carbon dioxide from the Decan Traps in India, or climate change resulting from continental drift as the culprit behind the mass extinction. Whatever the cause, almost all large vertebrates, on land, in the sea and in the air (dinosaurs, plesiosaurs and pterosaurs), along with most plankton and tropical invertebrates and about eighty percent of all species went extinct.

The survivors of the K-T event, as the catastrophe is called, were mostly small in size with modest needs, including mouse-sized mammals, birds and many small lizards (though large crocodiles somehow made it through the carnage). During the Paleocene Epoch (65 to 56 MYA), the first epoch of the Cenozoic Era, the weather remained as hot and humid as it had been in the Cretaceous. By the end of the epoch, the planet was once again covered in dense jungles and forests. The birds were the first land animals to take advantage of the absence of dinosaurs. Large flightless birds with giant heads and beaks, such as gastornis of Eurasia and the "terror birds" of South America, evolved quickly to fill the suddenly vacant ecological niche left by the carnivorous dinosaurs, to whom they were closely related. Mammals, however, remained small and peripheral. The vacancies in the sea left by the disappearance of mosasaurs and plesiosaurs were filled by sharks, which had been around for hundreds of millions of years, but which could now evolve into huge sizes.

Until the latter part of the Eocene (56 to 33 MYA), the climate was hot and humid, with rainforests stretching almost to the poles. Mammals began the epoch small in stature, but finally began to grow. By epoch's end, some reached stupendous size. Brontotherium, looking something like a rhinoceros, grew to 16 feet in length and weighed in at three tons. Predators grew in tandem with the herbivores, with the hyenalike andrewsarchus reaching six feet high at the shoulder, twelve feet long and weighing one thousand pounds. The first whales forsook dry land for the abundant feeding opportunities at sea. Some, like the basilosaurus, attained lengths of 60 feet and weighed up to 75 tons. As the planet cooled late in the Eocene, rainforests gave way to deciduous forests that could cope with the now seasonal temperature swings.

Probably the major innovation in the Oligocene Epoch (33 to 23 MYA) was the worldwide spread of newly evolved grasses. These lush fields of grass spurred the evolution of horses, deer and other ruminants and the carnivores that fed on them. Giant sharks ruled the seas, including the infamous Megalodon measuring fifty feet in length.

The final three epochs of the Cenozoic Era, the Miocene, the Pliocene and the Pleistocene, are germane to the matters addressed in Part III of this assessment, so we will halt our tour of life's unfolding at this point.

One further matter should be addressed before we leave the topic of animal evolution. The name of this subsection may be a bit misleading. Marchers know where they are going and set off with grim determination to get there. That's not, however, how evolution works. It is more like a hiker who doesn't know where he is or where he is going or when he will get there. He rambles hither and yon, with no particular goal in mind, simply following the easiest path. Because our travel through time ends with ourselves, the tendency is to see the history of life as a step-by-step progression toward ever increasing complexity and sophistication, with humanity as the apex of creation at the very tippy top of the tree of life. Humanity is certainly unique and wonderous, but we are more like an unexpected flower budding on a remote twig of a rambling, disorderly, overgrown shrub. When we are gone, life will continue on pretty much as it always has, and, in another 100 million years, we will look a whole lot less like evolution's final destination. Just bear in mind that there is no trajectory to evolution. There is no forethought or design. It is just a numbers game. Those organisms that develop traits which enable them to survive in the particular environment they find themselves will survive and reproduce, and those that don't won't.

4.   Analysis

In Part I, we were able to estimate the number of planets in this galaxy that are habitable for complex, intelligent life. We did so by analyzing the statistical likelihood of the various features that would be necessary for a planet to be considered habitable. Some of the numbers, such as the total number of stars in the galaxy or the percentage of each star type, are well supported while others, such as the number of stable solar systems, are more speculative. In most instances, though, we have at least some comparative examples upon which to base our estimates. Not so for life. We know of only one life, so we have no comparables. Consequently, we are not in a position to make a numerical estimate of the probability that any one of our habitable planets had, has or will develop life or the number of those that might develop complex life.

What we can do is examine life's history on this planet to see whether the emergence and the progression of life from simple to complex forms appears to have been natural, methodical and progressive, or whether it has involved elements of chance and serendipity. If the former, I think we can expect that most habitable planets will eventually develop complex life. If the latter, then only some percentage will.

Many theories (perhaps too many) have been advanced to explain how life emerged from the non-living. These stress that life self-assembled, beginning with simple, elemental processes and gradually evolving more complexity. The theories are quite reasonable, but the fact is that we really don't know how it all happened. More perplexing is the fact that we have only one example of life on this planet. All life, from oak trees to brewer's yeast, dinosaurs to bacteria, are DNA based, using the same fundamental chemical processes to live, and all are related, having descended from a single common ancestor. If the emergence of life is, in the proper environment, inevitable, it would seem there should be other trees of life on this planet that emerged independently from the non-living. But there are not. Perhaps there were others early on that were simply out competed by our life form, but that is only speculation. So, the emergence of life can be anywhere from inevitable to impossibly improbable. If the latter, the rest of this assessment is unnecessary, so we will assume the former.

Complex life requires oxygen and the eukaryotic cell. Oxygenic photosynthesis did not develop until more than a billion years after the emergence of life, and the eukaryotic cell another half billion years after that. Moreover, both happened only once, each within a single individual. There seems to be much that is accidental about the development of these fundamental requirements of complex life.

Then there is the matter of time. As we shall see later on, habitable planets will, over time, become uninhabitable. Complex life must arise before that happens. Yet, life seems to be in no hurry. The history of life on this planet is punctuated by long periods of stasis, where little or nothing happens. During the billion years before the development of photosynthesis, the half billion years before the emergence of the eukaryotic cell and the boring billion, life seemed content to just lay about, only doing enough to get by. Indeed, it was four billion years before the first animals of any size appeared on this planet, and, for much of the half billion years since then, they have just been doing the same things over and over again. And that's not to mention the several close calls with terrestrial and extraterrestrial calamities which could have wiped the slate clean of all complex life, leaving it to the protists and prokaryotes to try to start it up all over again.

I think we can reasonably conclude that the emergence of complex life on a habitable planet is not inevitable. Rather, it is difficult, uncertain and chancy. Thus, many, if not most, of our habitable planets will likely be microbial worlds bereft of complex life. And intelligent life? We'll address that in Part III.

III

INTELLIGENCE

In Part I of this assessment, we estimated the number of habitable planets there might be in this galaxy. In Part II we considered the chances that our habitable planets will harbor, not only life, but complex, multicellular animal life. In this Part III, we will examine the likelihood that any such complex life will lead to intelligent life.

There are many definitions of intelligence. Like the definitions of life that we have seen, most are descriptive, circular and of little help. Since most of us, I think, have a pretty clear understanding of what the term means, I won't bore you with the specifics. For our purposes, we will mean a life form that is capable of creating technology that enables it to communicate with life forms on other planets that are capable of doing the same.

1.   Brains

Human beings are vastly more intelligent than all other animals on this planet, or some of us at least. If intelligence is a function of the brain, what is it about our brains that make it so?

The most obvious reason is that our brains are so much larger than those of other animals. On average, the human brain has a volume of 1350 cubic centimeters and weighs 1400 grams (about 3.1 pounds). By comparison, the brain of a rat weighs about 2.0 grams (about 0.07 ounces); a domesticated cat 25 grams (less than an ounce); a beagle 72 grams (2.5 ounces); a pig 180 grams (6.3 ounces); a lion 240 grams (8.5 ounces); a cow 425 grams (15 ounces), a bear 450 grams (a pound); a hippopotamus 580 grams (a pound and a quarter); a giraffe 750 grams (1 pound, 10 ounces); a shark 32 grams (1.13 ounces); an alligator 8.4 grams (3/10ths of an ounce); an owl 2.2 grams (0.08 ounces); and a snake 0.1 grams (0.0035 ounces). That makes us look pretty smart.

On the other hand, there are 13 animals with bigger brains than ours, mostly varieties of whales, notably gray whales at about 4310 grams (nine and a half pounds); killer whales 5,620 grams (11.5 pounds); and the largest of them all, sperm whales at 7,800 grams (17.3 pounds). Then there is the bottle-nosed dolphin at 1500 grams (3.3 pounds) and the elephant at 4780 grams (10.5 pounds). But really, is an elephant more than three times as intelligent as a person, or a sperm whale six times? If so, that would be a dang smart whale.

You may have noticed that the larger animals tend to have larger brains. That makes sense, since larger creatures have larger parts. That's why they are larger. That caused scientists to wonder whether intelligence might be determined by the ratio between brain size and body mass rather than by brain size alone.

Using simple math, they found that the ratio of brain size (about 3.1 lbs.) to body mass (about 150 lbs.) in humans is about 1:50. So, what about those smarty pants whales and elephants? Well, it turns out that the brain/body weight ratio of dolphins is 1:83; of elephants 1:560; and of the sperm whale a whopping 1:5000. Human beings were back on top. The problem was that the brain/body weight ratio seemed to work well for larger animals, but it didn't do so well with smaller ones. The ratio for mice is 1:40, which is significantly higher than that for humans. Were mice truly twenty percent smarter than people? Worse yet, the ratio for wrens and other small birds was 1:12; for the tree shrew 1:10; and for ants, of all things, 1:7. Since no one ever says, "smart as an ant," the ratio method was clearly not a good indicator of intelligence. There is a good reason why it is not.

In 1762 the French naturalist, Albrecht von Haller, proposed that, though larger species have larger brains, brain size diminishes relative to body size in larger animals. Thus, brain to body size ratio will be higher in smaller animals than in larger animals. This is an example of allometric growth. Allometry is the study of the way body shape and the proportion of body parts will vary depending on body size. Allometry recognizes that some body parts will scale proportionately with an enlarged body, as is true with blood volume, which always remains a fixed proportion of body volume. Other parts, such as bones, will grow faster than the body as a whole, greater bone mass being necessary to support the load of the increased body weight. Still other parts will grow more slowly than the body as a whole, as in the case of the brain. In this way, the recognizable architecture of a class of animals can be maintained from its smallest to its largest member. If all parts scaled proportionately, it would be otherwise.

By way of example, if the brain of a sperm whale, which has a total body weight of 100,000 pounds, were scaled up proportionately from its cousin, the dolphin, whose brain/body ratio is 1:83, the sperm whale's brain would weigh in at twelve hundred pounds (i.e., 100,000 lbs. ÷ 83 = 1200 lbs.) rather than its actual seventeen pounds, severely distorting the whale's shape and likely making it dysfunctional and unrecognizable as a relative of the dolphin.

In his 1973 book, *Evolution of the Brain and Intelligence*, paleontologist Harry Jerison examined the allometrical relationship between brain size and body mass for the purpose of determining the intelligence of species of mammals. Jerison reasoned that animals needed a certain volume of brain to manage movement, sensory perception and other bodily functions. If a specie's brain size were larger than what was strictly necessary to manage bodily functions, such "extra" brain mass could be used for cognition.

Jerison examined a wide mix of species to calculate the allometric relationship between brain and body size in mammals. Using this calculation, he could predict the expected brain mass of a mammal of a given size (i.e., the amount of brains needed to maintain bodily functions). He could then compare the actual brain mass of a species with the expected brain mass of a species of that size to determine its "encephalization quotient" (EQ), that is, the amount, if any, by which actual brain size exceeded expected brain size. If the brain of a species was twice the expected size, that species was assigned an EQ of 2.0. If its brain was fifty percent larger than expected, its EQ was 1.5, and if it was exactly what was expected, its EQ was 1.0.

The EQ for humans, as it turns out, was 7.5, meaning that our brains were seven and a half times larger than what was necessary to maintain bodily function and indicating that we were by far the most intelligent of all species. Second to humans was the bottle nosed dolphin at 5.3. After that, various primate species came in with EQs of a little more than 2.0.

The encephalization quotient remained the most commonly used method of evaluating the intelligence of species for some three decades. But early in this century, its failings and deficiencies began to be exposed. For instance, the small capuchin monkeys were assigned an EQ of just over 2.0, the highest EQ of all non-human primates, while the EQ of the great apes fell below 1.0, even though the apes exhibit far more complex behaviors and abilities. Moreover, at an EQ of less than 1.0, the great apes, under this theory, didn't even have enough brain power to operate their own bodies ... a conclusion that doesn't square with the sight of orangutans swinging effortlessly through the jungle canopy.

Systematic studies of the cognitive abilities of mammal and bird species have since discredited the predictions made by EQ. In fact, these studies concluded that simple comparisons of absolute brain size are a more reliable measure of cognitive capabilities than the encephalization quotient. So, we were back to where we started.

There was another way, though. The brains of vertebrates are made up primarily of neurons (cells that receive and transmit information, and which are the working unit of the brain) and glial cells (which provide support functions for the neurons). In the early 2000s, one researcher, Suzana Herculano-Houzel, wondered if counting the total number of neurons in the brain would be the best way to objectively compare cognitive abilities among species. She soon found that there were no systematic counts of neurons in the brains of animals. It was, however, common knowledge among neuroscientists that the human brain contained, on average, 100 billion neurons. But even that number turned out to have been just an off the cuff estimate made by no one knows who that had been repeated so often that it had gained currency as undisputed truth. So, she would have to start from scratch. Incidentally, it had also been textbook fact that there were ten to fifty times as many glial cells in the brain as neurons, but, again, these numbers were apparently pulled out of thin air with no scientific support whatsoever. Actually, the ratio is closer to two to one. Beware of consensus science.

Counting billions of anything is, to say the least, tedious and time-consuming work. Thus, the first step was to devise a process for quick and accurate counts. After some trial and error, Herculano-Houzel developed such a process, which involved liquifying brains by dissolving cell membranes, then sampling and counting the cell nuclei floating freely in the brain soup.

Using this method, Herculano-Houzel established that the human brain contained, on average, 86 billion neurons. The approximate counts for other animals includes the gorilla 33.4B; orangutan 32.6B; chimpanzee 28.0B; baboon 10.9B; giraffe 10.7B; brown bear 9.58B; rhesus monkey 6.37B; lion 4.67B; dog 2.25B; pig 2.22B; cat 760M; squirrel 453M; rat 200M; crocodile 80M; and mouse 71M.

As you can see, the human brain contains far more neurons than any other animal. Well, except for one. Curiously, the elephant brain contains an amazing 257 billion neurons, three times more than humans. What was that all about?

In broad brush, the brain consists of three parts: the cerebral cortex, which controls higher functions, such as speech, thinking and memory; the cerebellum, which regulates motor movement; and the brain stem, which controls basic body functions such as breathing, swallowing and heart rate.

When Herculano-Houzel broke down her neuron counts into these parts of the brain, she found that the human brain's 86 billion neurons were comprised of 16B neurons in the cerebral cortex, 69B in the cerebellum and 1B in the brain stem. On the other hand, 98% of the 257 billion neurons in the elephant's brain are found in the cerebellum, leaving only a modest 5.6B neurons in the cerebral cortex. Apparently, the elephant's cerebellum is so large because it processes a vast amount of sensory information transmitted from its 200-pound trunk, which feeds directly into the cerebellum via the trigeminal nerve.

The cerebral cortex counts for rest of the above animals are: the gorilla 9.1B; orangutan 8.9B; chimpanzee 7.4B; baboon 2.8B; giraffe 1.73B; rhesus monkey 1.71B; squirrel monkey 1.34B; lion 545M; dog 530M; pig 425M; brown bear 251M; cat 250M; squirrel 71M; rat 31M; and mouse 14M. We have no actual counts of the numbers of neurons in the cerebral cortex of dolphins and whales, but it is estimated that they range from about 3 billion to less than 10 billion.

Since higher cognition is the function of the cerebral cortex and humans have far more neurons in their cerebral cortex than any other animal, we can now say conclusively that we are the most intelligent animal on the planet. Yay. But we sort of knew that already.

The really important take away from Herculano-Houzel's study, one which you might have already noticed, is that primates have far more neurons in their brains than other mammals having similar brain size. For example, a capybara (a large rodent species), which weighs over 100 pounds, has a cerebral cortex mass of 48.2 grams, while a bonnet monkey, which weighs nine pounds, has a cortex mass of an almost identical 48.3 grams. Yet, the capybara cortex contains 306 million neurons, while the bonnet monkey cortex contains 1.7 billion neurons. How could that be?

When Herculano-Houzel compared brain sizes and neuron counts among primates, she found there to be a linear relationship between the increase in brain size and the increase in the number of neurons. Thus, if the cerebral cortex of one primate species is twice as large as the cerebral cortex of another primate species, it will have twice the number of cortical neurons. That makes perfect sense. If one bag will hold a hundred pennies, a bag twice that size will hold 200 pennies. That is not, however, how it works for all other mammals.

The linear relationship between cortex size and number of cortical neurons translates (excuse the math) to an allometric exponent of +1.0. Non-primate mammal species, however, have an allometric exponent of +1.7, meaning that a tenfold increase in cortex size will result in not ten times as many neurons but only three times as many. Thus, the size of the cerebral cortex of non-primate mammals must expand hugely for even a little gain in the number of neurons.

The reason for this is actually quite simple. The size of the neurons themselves in non-primate mammals increases when the size of the cortex increases. Not so in primates, whose neurons stay the same size, even as the cortex grows larger.

Worse yet for the non-primates, the average size of non-primate neurons increases with the number of neurons in the cortex raised to the power of +0.6. Without getting too far out into the mathematical weed, this means that when the cortex of a non-primate gains 10 times more neurons, the neurons become 4 times as large, requiring the cortex to expand 10 x 4 or 40 times to accommodate the additional neurons (rather than just 10 times, as is the case with primates). When a non-primate's cortex gains 100 times as many neurons, its neurons become 4 x 4 = 16 times as large, requiring a cortex 16 x 100 or 1,600 times as large (rather than 100 times as large). With a thousand times more neurons, the neurons become 4 x 4 x 4 = 64 times as large, requiring the cortex to expand 64,000 times (rather than 1,000 times). The consequences of this non-linear expansion of neuron size can be illustrated by the following example. If a rat's brain were scaled up to the same 86B neurons that the human brain contains, its brain would have to weigh 57 pounds and its body 69 tons. What this means in practical terms is that non-primates simply cannot have a brain even remotely equivalent to the human brain.

We know that primates have far more neurons in the cerebral cortex than non-primates because primates have evolved neurons that do not get larger as the cortex expands. But what explains why humans have such an unusually high number of cortical neurons when compared to other primates?

Actually, nothing, because we do not. We seem to have just the right number of neurons for primates of our size. Using established primate neuron scaling, a generic primate brain containing 86B neurons should weigh 2.75 pounds with a body weight of 145 pounds, which is not far off from the actual human averages of a 3.1-pound brain and a 155-pound body. In other words, humans have a plain old, standard issue primate brain that, when scaled to our size, is essentially the same as all other primates ... all, that is, except the great apes.

Great apes are as large as or larger than humans (gorillas averaging about 350 pounds; orangutans close to 200 pounds; chimpanzees over 100). Yet their cerebral cortex contains only about half the number of neurons as the human cortex. It is not that humans have an unusually larger number of neurons but rather the great apes have an unusually low number. Their brains are way out of line with primate neuron scaling. Why is that? One word. Energy.

It is expensive to operate the brains of any animal, but it is most expensive for humans. While the brains of most animals consume from 2 to 10 percent of their energy intake, the human brain, which comprises only two percent of our total body mass, consumes 25 percent of the energy required for the entire body. The cost is a function of size.

If the brains of great apes were scaled up to their body size in the same proportions as other primates, they would have brains that are nearly as large as or larger than human brains. If that were so, their caloric requirements would be similar to our own. Apes are vegetarians, however. Such low-calorie fare requires them to spend about eight hours a day foraging for food. They spend another eight hours sleeping and the remaining eight hours taking care of ape business, whatever that might be. Researchers have found that eight hours a day is the practical limit to the time apes can actively forage. It takes all of that time to fulfill their energy needs, since, if they were taking in more calories than they needed, they would grow fat, and you don't really see a lot of pudgy, out of shape apes in the wild. If ape brains were as large as they should be for a primate of that size, their energy needs would soar well beyond their capacity to fulfill them. So, to maintain their large size, apes have had to evolve smaller, less energy hungry brains.

This recent research makes it clear that non-primates could never have attained a brain that would generate a level of intelligence comparable to our own and will never be able to do so in the future, absent a major evolutionary change in the structure of their neurons. Primates are a different story. Primates the size of the great apes will naturally develop brains as large as our own, provided the species is able to obtain sufficient energy resources. Since they cannot, their brains have remained small relative to body size. For their brains to expand to expected primate size, they would have to adopt an entirely new lifestyle. That is precisely what our ancestors did, as we will see in a moment.

2. Ancestors

375 million years ago, tiktaalik became the first fish to elbow its way onto dry (or, more properly, marshy) land. The first terrestrial vertebrates, the tetrapods, evolved from such land loving fish around twelve million years later. Amphibians made their appearance 340 MYA.

Amniotes evolved from amphibians 312 MYA. They differed from amphibians in that the embryo developed within a set of protective extra-embryonic membranes, that is, within an egg with a tough outer shell. The amniotic egg permitted amniotes to break their dependence on nearby water bodies for reproduction. Amniotes include all surviving reptiles, birds and mammals as well as their extinct relatives.

Only a couple of million years after their emergence, the amniotes split into two branches, the sauropsids and the synapsids. Descendants of the sauropsids include modern lizards, crocodiles, snakes, turtles and birds, along with a horde of extinct ancestors, including dinosaurs. The only surviving line of the synapsids is the mammals.

You can see that the line of animals from which mammals would spring diverged from the line from which reptiles, dinosaurs and birds would emerge not long after the first appearance of vertebrates on land. So, no, mammals did not evolve from reptiles, as it is often said. They are not some new and improved version of a primitive, out of date lizard. Mammals and reptiles parted ways three hundred million years ago and have developed separately and independently.

During the Permian Period (298 to 252 MYA), one group of synapsids, the pelycosaurs, became the dominant terrestrial animals, occupying most ecological niches and some obtaining great size. Midway through the period, they would be replaced by a more advanced from of synapsid, the therapsids.

The therapsids, often referred to as "mammal-like reptiles" (though they were not reptiles at all) differed from the pelycosaurs, the reptiles and the amphibians in that they held their limbs upright underneath their bodies rather than sprawling out to the sides. Also, their teeth diversified to suit their range of diets. They were far and away the most diverse and abundant large animals of the middle and late Permian and included a wide array of herbivores and carnivores, ranging from the size of a rat to large, bulky herbivores weighing a ton or more. Since the therapsids were the dominant terrestrial life form and mammals are their direct descendants, the Permian is sometimes referred to as the First Age of the Mammals.

Later, one group of therapsids would give rise to the cynodonts, a family of animals which began to develop characteristics that are today associated with mammals, such as warm blood, fur and complex, differentiated teeth. The larger therapsids did not survive the great dying that occurred at the end of the Permian Period. Fortunately for us, the cynodonts did make it into the Triassic Period (252 to 199 MYA), probably as a result of their mammal-like traits, small size and penchant for burrowing.

The other group of amniotes, the sauropsids, weathered the Permian extinction a little better than the synapsids. By the middle of the Triassic Period, they had diversified to fill the many ecological niches that had been left open by the mass extinction. One group, the archosaurs, became dominant (the name itself meaning "ruling reptile"). 230 million years ago, one lineage of the archosaurs evolved into dinosaurs. The landscape at the time was dominated by other archosaurs, and so the dinosaurs would remain small in size, number and diversification for the rest of the Triassic.

The first mammals evolved from one branch of the mammal-like cynodonts 210 million years ago (i.e., 20 million years after the first dinosaurs). As with the first dinosaurs, they would remain small and few in number throughout the remainder of the Triassic.

Another major mass extinction occurred at the end of the Triassic. The mammals managed to make it into the Jurassic Period (199 to 145 MYA), but the other cynodonts and therapsids did not. Now, mammals were the sole remining member of the synapsid branch of the amniotes.

Unfortunately for the mammals, the dinosaurs also survived the transition to the Jurassic. Most other archosaur species, however, died off, leaving the field wide open for the dinosaurs. They quickly grew and diversified. Soon they became and would remain the dominant terrestrial life form until the end of the Cretaceous Period (145 to 65 MYA), 135 million years later. Throughout this long period, mammals would remain diminutive. According to some, they amounted to nothing more than rodent-like creatures, feeding on plants and insects, living high up in the trees or in underground burrows, coming out only at night while the dinosaurs slept. The competition from the dinosaurs during the Jurassic and Cretaceous was fierce, literally, restricting the mammals to their dismal, little niche.

So, why was it that the dinosaurs were able to dominate terrestrial life to such an extent that mammals and their close kin would remain in their shadow for some 135 million years? Was it the fast start the dinosaurs got off to during the early part of the Jurassic? After all, they were a branch of the large-bodied Archosaurs which dominated much of the Triassic. Or did dinosaurs have some structural advantages? It might very well have been the latter.

Breathing for mammals is a matter of inflating and deflating the lungs (breathing in and breathing out). Air is breathed into the lungs, where oxygen is absorbed by surrounding blood vessels. At the same time, waste carbon dioxide is released from the blood vessels into the lungs and then exhaled. Dinosaurs breathed differently. The inhaled air was not immediately exhaled. Instead, the air was directed through an extensive system of air sacs throughout the body. This air sacs system gave their lungs a continuous, one directional airflow that enabled much higher concentrations of oxygen to enter the bloodstream.

Today, birds, which are a branch of the dinosaurs, utilize a very similar form of respiration. This highly efficient breathing method enables ostriches to run at top speed for up to thirty minutes (as opposed to only a few seconds for most mammals) and migratory fowl to fly over the Himalaya Mountains at altitudes where oxygen levels are extremely low. Throughout most of the Mesozoic Era, oxygen levels appear to have been very low, comprising perhaps 10 to 15% of air (as opposed to 21% today). The dinosaurs' efficient breathing system enabled them to remain active and grow to huge size even with such low levels of oxygen. Mammals, on the other hand, could have remained active only at small sizes where oxygen demands are lower.

The dinosaurs' method of respiration had another benefit as well. Small animals have a much greater surface area to volume ratio than do large animals. This makes it more difficult for larger animals to dissipate internal heat generated by their activities. The dinosaur's internal system of air sacs, however, took heat directly from the organs and expelled it during exhalation. This allowed dinosaurs to grow much larger than mammals ever could.

Another advantage the dinosaurs had was in their bones. Mammal bones are for the most part solid. The sheer weight of the bones of large mammals requires massive legs to bear it, ultimately limiting the size to which mammals can grow. The bones of dinosaurs, on the other hand, were pneumatized, meaning they had hollow spaces within them, as do bird bones today. Moreover, the bone that did exist was exceptionally dense, making their bones both sturdy and light. With less weight to support, dinosaurs could grow to immense sizes, far larger than the largest mammals. By way of comparison, African elephants, today's largest land mammal, can weigh up to 14,000 pounds and stand 13 feet at the shoulder. On the other hand, Argentinosaurus, the largest dinosaur, could weigh up to 200,000 pounds and reach lengths of 115 feet.

For these and other reasons, mammals were simply unable to compete with dinosaurs at sizes exceeding more than a few pounds.

Traditionally, Mesozoic mammals have been maligned as furtive, scurrying, rat-like vermin that eked out a meager existence in the shadow of the giants, doing little more than hiding out during their millions of years of subjugation. This was largely due to the paucity of the mammalian fossil record, which consisted mostly of tiny teeth and jaw fragments, and which compared so unfavorably to the spectacular dinosaur finds.

Since the late 1990s, however, there has been an explosion of early mammal fossil discoveries, particularly in China. The Chinese fossils were formed from volcanic eruptions that buried animals in ash, preserving them spectacularly. These fossils, including entire skeletons, with fur, skin and internal organs, have required paleontologist to reassess their prior notions.

Foreclosed by the dinosaurs from large size, Mesozoic mammals miniaturized. In this small world, they proliferated and diversified to create and fill a broad range of ecological niches. From digging and burrowing, to climbing into the treetops, to gliding (as do today's flying squirrels), to swimming beaver-like in freshwater ponds, mammal groups radiated and experimented throughout the time of the dinosaur. Indeed, their diversity and abundance back then is comparable to that of the small mammal fauna of today.

Being small, the Mesozoic mammals sought the safety of darkness. To do so, their senses had to evolve to enable them to navigate the nocturnal environment. For starters, light sensitive rods in their retinas proliferated at the expense of color sensitive cones. This allowed them to see well in the low light conditions of their burrows and the night, though rendering them colorblind, as are most of today's non-primate mammals.

Moreover, recent fossils show that the olfactory bulb within the brain, which processes smell, enlarged greatly during this time, endowing these early mammals with a keen sense of smell, far beyond that of other vertebrates.

Another innovation resulted from changes in dentation. To access diverse food sources, the Mesozoic mammals continued developing specialized teeth. More importantly, they learned to chew. Chewing, which is unique to mammals, requires teeth to move from side to side, in addition to up and down. Detaching certain small bones in the jaw created a more flexible joint for chewing. These small bones, as it turns out, were the malleus and incus, which were incorporated into the middle ear, along with a third bone, the stapes. This complex, three boned middle ear system is unique to mammals, and it enabled them to hear a wide range of high frequency sounds that were inaudible to other vertebrates.

Many early mammals were burrowers. In their underground dens, fur served as whiskers to provide them with tactile information regarding their surroundings. To process this information, their brains developed the neocortex, which is the center of higher brain function. It is also unique to mammalian brains. As the neocortex grew, so did the mammals' cognitive abilities.

It takes a lot of brain power to process sensory information, and these sensory innovations drove an explosion of brain size in the Mesozoic mammals. It also laid the groundwork for the evolution of ever bigger mammalian brains after the dinosaurs.

Another driver of larger brain size and cognition was socialization. Early in the Jurassic, many mammals switched from egg to live birth. Live birth results in fewer and less-developed young, requiring the adults to increase their investment of time and energy in their rearing. The longer time spent by mammal young under their parent's care fostered the development of complex behavior. At about the same time, mammals developed lactation. This allowed the young to be fed when the food consumed by the adults could not be digested by the young, enabling mammals to live in difficult environments. It also further increased the time spent by the mother in caring of her young. These lifestyle changes drove the development of a large forebrain which is associated with complex behavior.

So, the mammals did not lay idle during the Mesozoic as was previously supposed. Rather it was a time of significant experimentation and development of their defining characteristics, including, most importantly for our purposes, a trajectory toward higher intelligence. Be that as it may, they remained constrained by the dinosaurs to their small world for tens of millions of years.

Nothing lasts forever, however. The asteroid (or other causes) at the end of the Cretaceous laid the dinosaurs low, along with three quarters of all other species of animals on the planet. And so, one morning 65 million years ago, the mammals awoke for the first time in their 145-million-year existence into a world without monsters.

Their small size, taste for insects, ability to scavenge, penchant for burrowing, higher intelligence and the many other talents they had developed during their era in the small world had enabled the mammals to survive two mass extinctions, and their direct ascendant, the cynodonts, a third. Now it was their turn to surge. And so they did. If the Mesozoic Era (252 to 65 MYA) had been the "Age of the Dinosaurs," the Cenozoic Era (65 MYA to the present) would be the "Age of the Mammals." It should be borne in mind, however, that, absent the accidental, improbable collision 65 million years ago, there is no good reason to believe that dinosaurs would not have remained the dominant land animal to this day. Highly serendipitous.

Most Mesozoic mammal groups did not survive the end Cretaceous extinction event, but three did. Those were the egg-laying monotremes, the pouched marsupials and the placentals, which give birth to live offspring in an advanced stage of development. Monotremes today consist of echidna, platypus and spiny anteaters. Marsupials are a bit more numerous, with 250 species. Placentals, though, include more than 4,000 species.

There is ample evidence in the fossil record of marsupials and monotremes having been abundant possibly as far back as 175 MYA. There is, however, no definitive evidence of placentals until after the demise of the dinosaurs 65 MYA. Despite the absence of fossil evidence, DNA analyses have consistently suggested that, not only did placentals exist before that time, but a burst of placental diversification had occurred around 95 MYA, long before dinosaurs became extinct. That became common wisdom among paleontologists for a couple of decades. A recent study, however, re-examined the fossil calibrations of the prior studies to minimize potential biases in the molecular dating. The revised DNA sequencing showed that the major diversification of placental mammals occurred shortly after the extinction of the dinosaurs, a result that is consistent with the fossil record. That makes perfect sense, since, as we have seen, bursts of evolution require a permissive ecology, and the disappearance of the dinosaurs freed up a lot of ecological space into which the mammals could evolve.

The diversification of placental mammals during the early Cenozoic included the evolution of the Primate Order (of which, of course, we are a part). Primates are typically arboreal, dependent on sight rather than smell, and possess stereoscopic and good color vision. They also have larger brains than other mammals of equivalent size, probably because they are largely social animals that need large brains to navigate complex relationships within the group.

The early primates appeared by about 55 MYA. They were similar to squirrels and tree shrews in size and appearance and were probably nocturnal. Soon thereafter prosimians, such as modern lemurs and lorises, evolved. They developed grasping hands and feet, and the position of their eyes indicates they were developing more effective stereoscopic vision, enabling them to leap from limb to limb. Numerous genera of prosimians rapidly spread through North America, Asia, Europe and Africa. By 34 MYA, however, many prosimian species had become extinct, probably as a result of cooler temperatures and the appearance of the first monkeys.

Monkeys apparently evolved from prosimians around 37 MYA. Their fewer teeth, larger brains and more forward-looking eyes suggest that the early monkeys were becoming diurnal fruit and seed eating forest dwellers.

Apes diverged from the Old-World monkeys around 25 MYA. Apes are of two kinds, great apes (gorillas, chimpanzees, bonobos, orangutans and humans) and lesser apes (gibbons and siamangs). Our discussions will deal only with great apes. And that brings us to the Miocene Epoch (23 to 5.3 MYA), which is where we paused our whirlwind tour of life after the Cambrian in Part II.

The Miocene was the golden age of the great apes. Some forty genera of Miocene apes have been identified, and, considering the poor state of the fossil record for this period, there were likely many more. These apes adapted to a great variety of habitats, from tropical forests to semi tropical woodlands to, in some instances, scrubby brush. Their diets varied as well, with some apes specializing in leaves and others subsisting on fruit and nuts. Sizes ranged from the 10-pound East African micropithecus to the gigantic 600-pound gigantopithecus from southern China.

All fourteen or so genera of early Miocene apes inhabited Africa. As you might expect, these early apes retained many monkey-like features. A representative example of apes from this period is proconsul of western Kenya. Although proconsul was without a tail and had apelike dentation, it differed substantially from modern apes in mobility. As with monkeys, its arms and legs were nearly equal in length, and its arm movement was much more restricted than in modern apes and humans. Whereas today's apes travel through the trees suspended from the branches above them while holding their bodies vertically, Proconsul walked along branches on all fours with its body horizontal to the ground, as do monkeys.

The continents of Africa/Arabia and Eurasia had long been separated by the once but no longer Tethys Sea. About 17 MYA, the two continents converged, allowing for a great movement of animal species in both directions. Many African apes passed through Arabia and into Eurasia. The continental convergence coincided with what is called the Mid Miocene climactic optimum, a prolonged period of global warming (yes, global warming used to be thought of as a good thing). The warmer temperatures allowed lush tropical and subtropical forests to extend northward. A continuous supply of ripe fruit and easy arboreal passage enabled the great apes to spread across southern Eurasia, from western Europe to China, during the period from 13 to 9 MYA.

One of the most studied apes of the Mid-Miocene is dryopithecus. This chimp sized ape was abundant throughout western and central Europe. Its arms display the same range of movement as modern apes and humans. This and many other adaptations to suspension suggest that dryopithecus travelled through the forest canopy in much the same way as today's apes.

During this same period, the climate of Africa became dryer and more seasonal. The scarcity of fossil evidence of apes in Africa between 16 and 10 MYA has given rise to the "out of and back into Africa" theory, which proposes that hominids (modern great apes and their immediate ancestors) arose in Eurasia, but 9 MYA, one lineage returned to Africa to give rise to modern apes and humans. The theory is a long way from being proven, but, if correct, dryopithecus would be a good candidate for being, as they say, our great grand ape.

In the late Miocene, Alpine, Himalayan and East African mountain building combined with shifting ocean currents to bring the Mid Miocene optimum to an end. Temperate, seasonal climates returned to Europe, replacing subtropical, evergreen forests with deciduous woodlands. By 5 MYA, Eurasian apes had been reduced to their present range of southern China and Southeast Asia.

Our story will continue with the surviving great apes in Africa.

3.   First Steps

As we have seen, adaptive radiation occurs when, because of a natural disaster or climate change, many species suddenly die off. The abundance of newly unoccupied ecological niches allows for mutations that would not have otherwise survived to flourish. Whatever can survive will survive, at least until populations increase again and natural selection starts weeding out the less fit.

Adaptive radiation will also occur when a species evolves a new feature that permits its members to exploit the existing ecology in a novel way. A burst of evolution follows as the species fills new niches. That is what happened 4.2 million years ago, when our ancestor, Australopithecus, came down from the trees and walked the Earth.

Radiometric dating tells us that the orangutan and the rest of the living great apes diverged from a common ancestor 18 MYA; gorillas and the chimp/human line split 10 MYA; chimps and humans 7 MYA.

We do not know what the last common ancestor (LCA) of chimps and humans looked like, though it would be natural to assume that it looked more like a chimp than like us, since chimps seem to us to look more apelike than do humans. Interestingly, however, Ardipithecus (4.4 MYA), the presumptive immediate ancestor of human's immediate ancestor, the Australopithecines, apparently travelled along tree branches on all fours as do monkeys and as did early apes, like Proconsul. If that is the case, it is probable that the LCA did so as well, since it is much more likely that chimps would have subsequently evolved all of the intricate adaptations necessary for suspensory locomotion than for Ardipithecus to have lost them.

Ardipithecus possessed certain adaptations for walking upright, such as the spinal cord entering through the center of its skull (like humans) rather than the back of the skull (as with other apes) and an arched foot. It retained, however, many apelike adaptations for climbing, such as opposable big toes, which would have made walking upright awkward.

The serious business of bipedal walking began with the Australopithecus. The earliest known species of this genus or group is Australopithecus Anamensis (or just A. Anamensis)

(4.2 to 3.8 MYA). They were followed by A. Afarensis (3.9 to 2.9 MYA), a group that includes the famous Lucy, whose 3.2 million years old relatively complete skeleton was discovered in Ethiopia in 1974. A. Afarensis was partially contemporary with A. Africanus (3.3 to 2.1 MYA), but, whereas A. Anamensis and A. Afarensis lived in east Africa, A. Africanus resided in southern Africa. Then there is A. Garhi (2.5 MYA) discovered in Ethiopia, and A. Sediba (1.98 MYA) found in South Africa. There is also a so-called robust group of Australopithecus having large, thickly enameled cheek teeth and powerful jaw muscles ideal for chewing coarse, vegetal matter, which are collectively referred to as Paranthropus Robustus (2.3 to 1.2 MYA).

Whether this abundance of Australopithecines were just local adaptations of one or two species or were each a separate species is much argued but difficult to say with any degree of certainty in view of the sparse fossil record from that distant time. It has been said that "(t)he total world archive of hominid and early human bones could fit into the back of a pickup truck." The appearance, however, of numerous species to take advantage of the new opportunities presented by the Australopithecines' new lifestyle would be consistent with the idea of adaptive radiation.

Despite their many variations, the Australopithecines shared a number of fundamental traits.

Walking. The most obvious of these shared characteristics is walking. All Australopithecines walked and walked well. Other animals do or did walk bipedally, such as kangaroos and therapod dinosaurs, such as T-Rex. They, however, walk or walked with their backs nearly horizontal to the ground, using their tails as counterweights. Since apes have no tails, another method had to be devised. That method was upright, bipedal walking. It is a strange form of locomotion that no animal, other than humans and our immediate ancestors, have ever used. And for good reason. It's hard to do.

Upright, bipedal walking is a complex maneuver. It begins with the swing phase, where the leg pushes off the ground from the big toe, swings under the body while slightly flexed, then straightens out prior to the foot hitting the ground, heel first, and is followed by the stance phase, where the leg remains in the extended position to provide support for the other leg as it goes through the swing phase. This requires balance, coordination and split-second timing. It made extensive anatomical adaptations necessary as well, including a curved lower spine which allows for upright posture; a broad pelvis that efficiently distributes the stresses generated by upright posture; thigh bones that angle inward so they can swing below the body's center of gravity; long lower limbs; the ability to fully straighten the leg; an enlarged, in-line big toe; and a foramen magnum (the hole where the spinal cord enters the skull) that enters the center of the skull rather than the rear, as with present day apes (otherwise our heads would be pointing upward). Moreover, a locking knee joint is necessary for standing for long periods without straining the leg muscles.

Australopithecus had all of these adaptations, which strongly implies that it walked much as we do. No one, of course, has actually seen an Australopithecus, so there were some doubts as to how effectively they walked. In 1976, however, paleontologists working at the Laetoli site in the Tanzanian portion of the Rift Valley discovered animal trackways that had been laid down 3.6 million years ago on a layer of wet volcanic ash that subsequently dried and hardened. The tracks included a footprint trail some 80 feet long made by a pair of Australopithecines. The prints are almost indistinguishable from modern human footprints made on a wet beach. The shape and depth of the prints and the distances between them leave little doubt that Australopithecus was, indeed, an upright, bipedal walker.

The question is why. Why would a group of apes leave the trees and abandon a lifestyle that had served them well for more than 18 million years? The answer, once again, is climate change. For the most part, Australopithecus lived during the Pliocene Epoch (5.333 to 2.5 MYA), which followed the Miocene. During the Pliocene, the cool, dry seasonable weather that had begun in the late Miocene continued, fracturing the tropical forest into a mosaic of forests and dry grasslands. Deep forests remained, providing habitat for chimps and other apes, but, out on the fringes, the apes began experiencing a loss of habitat and the food resources that go along with it. There were resources out on the savanna, but the apes would have to come down out of the trees to access them. It is not unusual for chimpanzees and other apes to leave the trees, but it is usually only for limited purposes. To actually make your living on the ground required a more efficient type of locomotion than the waddling and knuckle walking practiced by chimps and gorillas. Bipedal walking fulfilled that need.

By way of example, the Laetoli footprints were laid down in what was then an open, flat plain. From the direction of the trail, the pair of Australopithecines were undoubtedly headed for the Olduvai Basin, which then consisted of a forest surrounding a shallow lake, and which would have been clearly visible a few miles ahead. The journey would not have been possible for a chimp. For Australopithecus, it would have been a walk in the park. Being able to travel long distances would have enabled the Australopithecines to access and exploit new habitats, which, in hard times, could mean the difference between life and death.

Another important advantage of bipedal walking is that it frees the hands. Free hands would allow an Australopithecus bread winner to bring food gathered in the grasslands back to the family. More importantly, free hands would have enabled Australopithecus to carry and use sticks. If you have ever had a pleasant walk interrupted by the sudden appearance of a large, snarling dog, you realize just how defenseless our kind is in such confrontations. A baseball bat would help even the odds. Similarly, an African savanna a couple of million years ago could be a dangerous place. A means of defense is always a good idea. Moreover, a stick would be useful in killing snakes and lizards and other small animals that might make a nice meal but which would be dangerous to catch by hand. Moreover, even today simple sticks are used by primitive peoples to dig up roots, tubers and other nutritious treats underground.

Other advantages of walking upright have been proposed, such as being able to peer over tall grasses or limiting one's exposure to the tropical sun by standing vertically rather than horizontally. I doubt such traits by themselves would have been sufficient to stimulate the anatomical changes required for bipedal walking, but they probably would have conferred some additional survival advantage, which would have further hastened the transition to walking.

Climbing. Below the waste, Australopithecus would have looked similar to humans, except that its toes were longer and its legs were proportionately shorter and covered in fur. Above the waste, not so much. Its face was apelike, with a low forehead, prominent brow ridges and a protruding lower jaw. More importantly, its arms were long and powerful, and its rib cage tapered upward from a broad base, such that its shoulder joints were closely spaced. Its hands were shorter than those of most apes, but it retained long, curved fingers with strong flexor tendons, indicating a powerful grasping ability. All of these features facilitated climbing and show that Australopithecus was as at home in the trees as it was on the ground.

There were certainly good reasons for retaining its climbing ability. There were still fruit and other forest resources in the trees. Moreover, the grasslands were full of dangerous predators. Having the ability to scurry up a tree upon one's approach would have been highly beneficial. Also, it is likely that Australopithecus slept at night up in the trees, safe from marauding nocturnal predators. Walking on two feet might be far more efficient than knuckle walking, as do gorillas and chimpanzees, but it is far less efficient than walking on all fours. The survival advantages of retaining the adaptations to climb may explain why Australopithecus became bipeds rather than quadrupeds.

It is important to bear in mind that the ability to climb was not some ancestral holdover in the transition from an arboreal to a terrestrial way of life. Instead, Australopithecus continued this hybrid lifestyle throughout its long tenure on this planet. They were and remained bipedal apes.

Size. Australopithecus was rather small, roughly chimp size, with males measuring about 4'6" inches in height and weighing 90 pounds and females about 3'6" inches and 65 pounds. They stayed roughly this size throughout their entire existence, likely remaining small so they could climb effectively, since climbing gets more difficult with size.

As you may have noticed, male Australopithecines were substantially larger than females. Physical differences in a species between the sexes is referred to as sexual dimorphism. Sexual dimorphism as it relates to body size is common among primates.

There is a strong correlation in primates between mating strategies and sexual dimorphism. In monogamous species, such as gibbons, male-on-male aggression is limited, resulting in little difference in size between males and females.

In polygamous species, males must compete against each other for access to females. One way to do so is to grow larger. Thus, in harem species, such as gorillas, where inter-male aggression is fierce, males are nearly twice the size of females. Among the Australopithecines, males were about 35% larger than females. Does that indicate a high level of competition among males for the privilege of mating?

There are, though, reasons other than mating that might cause males to grow larger than females. Chimpanzees live in promiscuous societies, yet males are about 35% larger than females. Humans are monogamous (at least some of them), but males are still about 20% larger. Chimpanzees, however, form bands to patrol their territories and fend off intruding males from neighboring groups, and early humans were in constant conflict with surrounding tribes. Such territorial, male-on male aggression may have been the stimulus for male growth.

Another, and perhaps even more important, sexually dimorphic trait among primates relating to mating strategies is the size of male upper canine teeth. Almost all primates have large, sharp, self-honing upper canines. When competing for mates, males display these daggerlike teeth and use them if necessary. As with humans, the canines of Australopithecus exhibit little sexual dimorphism. That would indicate less aggressive competition for mates amongst male Australopithecines, despite the considerable size difference between the sexes.

Brain Size. The brain of Australopithecus was small in comparison to those of humans, but about what you would expect of an ape of that size. Their cranial capacity ranged from about 380 to 485 cubic centimeters. By way of comparison, chimpanzee cranial capacity runs from 280 to 450 cc, while humans average about 1350 cc.

Australopithecus brain size remained stable until about 2.1 MYA with the appearance of Homo Habilis (handyman) (or, more properly, Australopithecus Habilis) (2.1 to 1.5 MYA). At 550 to 600 cc, however, the brains of A. Habilis are only slightly larger than those of other varieties of Australopithecus. It does indicate, though, the beginnings of a trend toward larger brains. Incidentally, Habilis was originally classed in the genus Homo (man) by its discoverer, Louis Leakey, because of its association with stone tools, which he believed could only have been made by man. Otherwise, it appears to be only a variant of Australopithecus and should be considered as such.

Diet. Like today's apes, the predecessors of Australopithecus ate almost exclusively forest fruit. As the forest diminished and the grasslands encroached, Australopithecus was forced to supplement its diet with foods from the savannah, which included nuts, seeds, tubers and roots. They developed large, thick-walled, flat molars capable of dealing with such tough and brittle foods.

In view of their large molars and only moderate sized incisors, it was long believed the Australopithecines were strictly vegetarian. Recent studies suggest otherwise, however. Carbon has two stable isotopes, C_{12} and C_{13}, both of which are absorbed by plants from the atmosphere during photosynthesis. There are two main photosynthetic processes, the C3 process, used by trees, bushes and shrubs, and the C4 process, employed by grasses and sedges. The C4 process takes up more C_{13} than does the C3 process, meaning that grassland plants will contain more C_{13} than forest plants. Since animals incorporate carbon from their food into their bones and teeth, an analysis of bones and tooth enamel can indicate whether the animal's food came from the forest or the savannah. Tooth enamel from Australopithecus shows that 25 to 50 percent of its food was derived from grasses and sedges, indicating they were eating not grass but the hyraxes, antelopes and other grazing animals that did.

It appears, then, that Australopithecus ate fruit, nuts, seeds, tubers, roots, bulbs, reptiles, grubs, small mammals, grazing animals and about anything else that came along. They had become flexible omnivores, a talent that enabled them to thrive in diverse environments and survive a long period of climate change, an ability they would bequeath to us.

Tools. For a long time, it was believed that only human beings had the capacity to make and use tools... "man the toolmaker", "it's what separates man from beast", and all of that. Then in the 1960s, the primatologist, Jane Goodall, while working in Tanzania reported that the chimpanzees she was studying would occasionally snap a twig from a bush, strip off the leaves and then stick the twig into an opening in a termite mound. After a few seconds, the chimp would withdraw the twig, to which termites were now clinging, and scrape them off into its mouth with its teeth, for a presumably tasty snack. Other researchers observed chimps using rocks to crack open nuts or leaves to sponge up fresh water from deep holes. In one area, chimps would from time to time break off a small tree limb, scrape off the twigs and bark, sharpen one end with its teeth then plunge the sharpened stick into the kind of deep hole in a tree where bush babies (a small nocturnal primate) typically sleep during the day, most often with no result. Occasionally, however, a chimp would manage to skewer a hapless primate, which it would devour on the spot.

Admittedly, twigs, rocks, handfuls of leaves and sharpened sticks are on the low end of the technology spectrum, but they are, in fact, tools. What's more, this behavior is not instinctive. Younger chimps learn how to make and use tools by observing older chimps then trying it out themselves, effecting a kind of intergenerational cultural transmission. Moreover, different chimps in different areas use different kinds of tools in different ways. Chimpanzee tool use, then, is a learned behavior that differs from human tool use only in degree.

Australopithecus was as or perhaps slightly more intelligent than chimpanzees, and, therefore, it is almost certain they had the ability to use and make simple tools. Chimpanzees typically use and discard tools after a single use, since carrying around a tool is burdensome and impractical for climbers and knuckle walkers. Furthermore, the types of food they access by their use of tools is not a part of the chimp's regular diet, rather more of a treat or diversion. Australopithecus, on the other hand, with their free hands, could easily carry sharpened sticks, clubs or other instrumentalities around with them, giving them the opportunity to explore possible uses of and make improvements to their tools. Unlike chimps, the types of food that these tools would allow Australopithecus to access would likely have been staples of their diet. Clubs and spears would allow them to capture small animals, and sharpened digging sticks would enable them to dig up roots, bulbs, tubers and burrowing animals, just as modern humans do today in parts of Africa.

Moreover, Australopithecus spent much if not most of its time on the savannah. Other savannah primates, such as baboons, are endowed with long, razor sharp canine teeth with which to discourage attacks by predators. Not so for Australopithecus, which had no natural defenses, other than, you know, kicking or slapping, which, I submit, would have been largely ineffective against a dangerous predator like a leopard. It would have made good sense for them to have armed themselves with proto spears and clubs for defensive purposes, and there is a strong belief among many paleontologists that they did.

Sticks do not last long in the tropical heat and humidity of Africa, so we will never know for certain whether Australopithecus actually did use such tools. Stone tools are another matter. They last virtually forever. So, if Australopithecus used stone tools, we should be able to find evidence that they did.

In fact, such evidence has been found. At the Gona site in Ethiopia, simple stone tools have been found that date back to as many as 2.6 million years ago. These consist of small cores or cobbles of volcanic rock that had been hit with another rock (referred to as a "hammer stone") to detach sharp flakes an inch or two long that could be used for cutting. The remaining cores could then be used to smash open objects, such as bones to extract marrow and skulls to get to the brains. Such tools are referred as "Oldowan" technology, from the Olduvai Gorge where such tools were first discovered. Oldowan tools are considered by paleontologists as Mode or Level 1 tools, that is, the first of five generations of stone tools. Although simple, they are highly effective, as archeologist have shown by butchering an entire elephant using such sharp flakes. Mode 1 tools continued to be used for nearly one million years.

In 2010, at the Dikika site in Ethiopia, researchers found bone fragments from a cow-sized mammal and a goat-sized mammal dating back to 3.4 million years ago. Closer inspection using electron imaging and x-ray spectrometry revealed cut-marks of the kind that can only be produced by stone tools. The bones were found only two hundred meters away from the site where an A. Afarensis skeleton had been unearthed, making it clear that Australopithecines were using sharp-edged stone tools to carve meat from bone some 800,000 years before the Gona tools were made. There was one problem with the discovery... no cobbles or core or flakes were found at the site. This has caused some scientists to question whether the cut marks on the bones really had been made by stone tools.

More likely, Australopithecines were simply using naturally occurring stone fragments. Such fragments might not have had the razor-sharp edges of manufactured flakes, but they would have been adequate to do the job. This find is strong evidence that Australopithecus was using, if not making, stone tools by this early date.

Hands. Human hands have broad palms, long thumbs and the ability to oppose the thumb to the tips of all of the other fingers. These features enable us to precisely manipulate objects. The hands of modern apes are long and narrow with muscles and tendon that generate a powerful grip, which is a highly desirable trait for life in the trees. Ape hands, though, are poorly designed for tool use.

So, what about Australopithecus? In 2015, researchers with the Max Plank Institute for Evolutionary Anthropology studied how fossil species used their hands by examining the internal spongey structure of bone called trabeculae. An initial examination of the trabeculae of chimpanzee hand bones showed that chimps simply cannot assume the postures necessary for tool use. An examination of the hand of an A. Afarensis, however, revealed that Australopithecines had a human-like trabecular pattern in the bones of their thumbs and palms that was consistent with the forceful opposition of the thumb and fingers necessary for tool use. This is further evidence that Australopithecus was using tools more than three million years ago.

Interestingly, it appears that the ability of Australopithecus to use tools evolved in tandem with refinements in their hands, more dexterous hands giving rise to more sophisticated tools, which then stimulated further improvements in the hands, and so forth.

Hunting/Scavenging. We have evidence that Australopithecus was butchering large savannah mammals by at least 2.6 MYA and probably by 3.4 MYA. The question is whether they hunted and killed the beasts themselves or just scavenged another animal's kill.

The only weapons Australopithecus may have ever had were sharpened sticks and clubs. I'm guessing the likelihood of their bringing down a fleet footed antelope with one of those is about equal to your chances of bagging a big buck with a baseball bat. Scavenging is another possibility. Dead carcasses, however, pose their own set of problems. Once decomposition begins, which under the tropical sun is pretty quick, virial, microbial and parasite populations surge, emitting odors and gases and making the carcass toxic and unappetizing. With no buzzard style specializations for such a lifestyle, Australopithecines would probably have been no more inclined to dine on a rotten corpse than any one of us would.

A possible alternative is meat stealing. Leopards often stash their kill high up in a tree safely away from lions and other terrestrial carnivores. With their climbing abilities, Australopithecines could have sneaked up a tree while the cat was away and taken all or a hunk of the kill, though a bold enterprise that would have been, given that Australopithecus was a regular feature on the leopard's menu. More plausibly, the Australopithecines could have sallied forth armed and in force, pelting the predator with rocks (which, it appears, they could have done) and chasing it from its prey. I imagine even today a dozen or so hardy lads brandishing spears and clubs, chunking rocks and hollering for all they were worth could likely drive a large male lion from its kill. Maybe could. Possibly could. If any intrepid reader would like to give that a try, I ask that you, or your next of kin, let me know how things works out.

Whatever they did, it does appear that, over time, Australopithecines were adding more meat to their diet, which would have significant long-term consequences.

Assessment. Australopithecus set the stage for those who would follow. They came down from the trees and learned to walk bipedally and upright. They developed a generalist, omnivorous lifestyle. They evolved precise, intricate hands and used them to make tools and weapons. They began to hunt (or scavenge), obtaining the additional nutrition necessary for their brains to grow. In a very real sense, we have only elaborated on what they initiated.

We should, however, avoid the inclination to see Australopithecus as merely a steppingstone to humanity. Australopithecus was a highly successful species (or group of species) in its own right, surviving severe climate and environmental changes, spreading across eastern and southern Africa and persisting for some three million years. Yes, they are gone now, but someday so too will we. Perhaps we will be followed by more advanced and intelligent beings, whether our own descendants or some form of AI. Maybe they will draw their own tree of life with themselves at the very top and us a step below. But I must say that on a cold, blustery winter's evening when I settle down in a comfortable chair in front of a cozy fire with a snifter of fine Armagnac and a good book in hand, I don't feel much like a steppingstone. I don't think Australopithecus did either.

4.   Ice Ages

When was the ice age? Look out the window. We are living in an ice age that has held the Earth in its grip for the past 2.6 million years. This most recent of the planet's ice ages, referred to as the Pleistocene Epoch (2.6 MYA to the present), is a natural continuation of the falling temperatures that began shortly after the age of the dinosaurs ended 65 million years ago.

The Pleistocene began with the formation of the polar ice caps, which have periodically expanded and contracted over the past 2.6 million years. During this time, there have been numerous long periods of glaciation, followed by short warming periods. At times, up to thirty percent of the Earth's landmass was covered in snow and ice, with glaciers building up in some places more than two miles thick and extending down Europe as far as 40 degrees north latitude. These vast ice sheets tied up a large part of the planet's water, causing sea levels to drop as much as four hundred feet below today's levels. With so much water tied up in glaciers, most of the world became very dry, with deserts emerging and expanding. During cold periods, the Sahara extended over a much larger area than it does today. Tropical forests of Africa were largely replaced by open woodlands, with jungle found only along rivers and humid coastal areas.

The early Pleistocene was cold but not bitterly so. Glaciers were small, and, in Europe, limited to northern Sweden and the Norwegian mountains. The cycles between cold and warm periods averaged about 41,000 years. By 800,000 years ago, however, the cold became severe. Except during the warm interglacial periods, southern Scandinavia and northern Germany (not to mention much of North America) were covered by ice. During that time, the glacial cycles changed to 100,000 years, averaging 90,000 years of cold followed by 10,000-year warm, interglacial periods.

There are a number of factors that contributed to the continuous oscillations between ice and warmth, such as the position of the continents and the paths of ocean currents. The driving force, however, has likely been Milankovitch

Cycles. During the first world war, a Serbian engineer, Milutin Milankovitch, theorized that the colder and warmer periods during the ice ages were due to small cyclical variations in the Earth's orbit around the Sun and in its rotation on its axis. By then, it was well known that the gravitational forces of the Sun, Moon, Venus and Jupiter caused three separate perturbations of the Earth's orbit and rotation: axis tilt, which varies between 22.1 degrees and 24.5 degrees over a period of 41,000 years; orbital eccentricity, being the variance of the Earth's orbit from near perfect circularity to its elliptical maximum and back again over a 100,000 year period; and precession, which is the wobble in the Earth's rotation, which takes 26,000 years to come full circle. Each of these orbital and rotational variations affects the amount of solar energy received over a given interval by any particular place on Earth. According to Milankovitch, the determining factor in glaciation is the amount of solar energy received during the month of June at 65 degrees north latitude (north because that is where most of the land is). He reasoned that, if the solar heat at that latitude in the month of June is not sufficient to melt the winter's continental ice and snow, the ice will accumulate over the years, eventually forming glaciers. Though not without question, Milankovitch's theory has today gained wide acceptance because observed temperature variations through the Earth's history agree so well with those predicted by the theory.

We are now living in the Holocene Epoch, a warm period which began with the end of the most recent period of glaciation 11,700 years ago, if we count from the end of the Younger Dryas cooling period (which we will get into later) or 14,700 years ago, if we count from the beginning of the Bolling Allerod warming period (which we will discuss later as well).

The Holocene was preceded by Wurm Glacial (115,000 – 11,700 years ago). The early Wurm was relatively mild and punctuated by warm interstadial periods. Glaciers were limited to Scandinavia and northern Europe. Around 74,000 years ago, however, Europe entered a deep and prolonged freeze. Glaciers covered most of Scandinavia, Britain and Ireland, and steppe-tundra covered most of northern and central Europe. This intense glacial period lasted until 59,000 years ago, at which point the climate became less severe. Then, 30,000 years ago, the world began to grow very cold once again, with the Last Glacial Maximum (i.e., the maximum extent of glaciation during the Wurm period) occurring between 22,000 and 19,000 years ago. Once again, much of northern and western Europe was ice burdened, and southern and eastern Europe became frigid tundra where the mammoths and the wooly rhinos played. During the Bolling-Allerod period (14,700 – 12,900 years ago), the world rapidly warmed. The glaciers began to melt, and temperate forests expanded. Then, 12,900 years ago, a sudden cold snap referred to as the Younger Dryas sent temperature plummeting to near glacial levels. Twelve hundred years later, the Younger Dryas ended as abruptly as it had begun, and temperatures started to rise to modern levels.

The Wurm Glacial was preceded by the Eemian Interglacial, lasting from 130,000 to 115,000 years ago. For a substantial part of the period, temperatures were significantly higher than they are today. Rhinoceros, elephants, lions and hyenas roamed northwestern and central Europe, while hippos and crocodiles basked along the banks of the Thames and Rhine Rivers.

The Eemian Interglacial was preceded by the Riss Glacial (300,000 to 130,000 years ago, which actually includes two separate ice events and an intergalcial). The Riss was preceded by the Holstein Interglacial, which preceded the severe Elster Glacial, and so forth and so on, long glacial periods followed by short interglacials, for 2.6 million years.

Since there is no good reason to believe that the ice will not return one day, the Holocene is merely one more interglacial period. When will the ice return? Hard to say. The present interglacial period has lasted 12,000 or perhaps 15,000 years, which is already longer than the average of 10,000 years. The Younger Dryas cold snap took only a few decades to return a warming Earth back into the deep freeze, so it could happen at any time and quite suddenly. Some scientists, relying on Milankovitch's theory, however, have calculated that the Earth should remain warm for another 50,000 years, though others are doubtful. Claiming that the 2.6-million-year pattern of short interglacials is changing just as we arrive on the scene sounds to me a bit like wishful thinking. The truth is that we do not know when the ice will return, and there is nothing we can do about it, anyway.

The Pleistocene has been a time of unsettled environmental conditions. In Europe, the comings and goings of the glaciers could rapidly alter the landscape from tundra to pine forest to broadleaf forest, each with its own fauna, and back again. The climate remained relatively warm in Africa, but rainfall varied dramatically, often over short periods of time. There were long periods of drought, tropical forests and deserts expanded and contracted and local environments could change from desiccated scrubland to humid, lake pocked river basin, and vice versa, seemingly overnight. It was in this welter of changing environmental conditions that our genus, Homo (man), was born and grew up.

Specialist species evolve special adaptations with which to exploit its particular environment. That's fine, until the climate changes. Generalist species, on the other hand, are less capable of exploiting a particular environment but are adaptable to a variety of environments, making climate change survivable. Humans, and the Australopithecines before them, have been the ultimate generalists. Having no particular specialization, our genus has relied on its steadily increasing intelligence and the culture generated thereby to cope with whatever has come its way. Improvise, adapt and overcome, as the slogan goes.

Such adaptability to a constantly changing environment has allowed our genus to evolve much more significantly and rapidly than any other group of mammals over the same period. As we shall see next, our group of tropical apes was able to thrive and multiply in the unsettled, often harsh ice age conditions.

5.   Homo

1.9 million years ago, the first species of our genus Homo (man, or, if you prefer, human) appeared in east Africa. It has been given the name Homo Ergaster (Working Man).

Paleontologists have been aware of Ergaster since the 1975 discovery of a cranium in 1.8-million-year old sediment near Lake Turkana in Ethiopia. The cranium did not resemble anything known from earlier times. The face did not strongly project, as does an ape's, and it would have had a protruding nose, strikingly departing from the flattened mid-face of apes and Australopithecus. Moreover, its cranial vault would have contained a brain weighing in at 850 cc, significantly larger than the brains of even the most advanced of the Australopithecines, A. Habilis, at 600 cc. It was clear that, by 1.8 MYA, an entirely new type of hominin (species in the human lineage) was on the scene.

Just how different Ergaster was from what had come before was not fully appreciated until 1984, with the discovery of an almost complete Ergaster skeleton near Lake Turkana that dated to about 1.6 million years ago. Since the individual was an adolescent at the time of his death, he has been referred to as "Turkana Boy." Boy, who at first was believed to be about 13 years old, stood 5'3". Because of his youth, it was thought that he would exceed six feet by maturity. Further study, however, indicates that he was actually only 8 or 9 years old at death. Like apes, Ergaster grew up quickly, with only a short interval of adolescence separating the child from the adult. Thus, at 8, Boy had attained the physical development level of a 13-year-old modern human. It is now thought he would have grown only another three or four inches. But even at 5'6", he would have been considerably taller than Australopithecus, which ranged up to a maximum of about 4'10".

Although significantly different from the apes and Australopithecus, the Boy's head, and those of his fellow Ergaster, was still primitive in many respects, with an elongated braincase, flat, receding forehead and a prominent brow ridge over his eyes. From the neck down, however, the Ergaster body plan was very similar to our own and significantly different from the bipedal apes. Its legs were long, and its arm to leg proportion was similar to that of modern humans. It had the barrel shaped chest of modern humans, rather than the funnel-shape of apes and Australopithecus. The scapular spine of its shoulders pointed downward like ours, which is useful for stone tool making and high-speed throwing, rather than upward like an ape's, which is necessary for tree climbing and swinging through the branches. As shown by 1.5-million-year-old preserved footprints (second oldest to the Laetoli prints), its feet were arched with inline big toes and were essentially modern in structure.

Here, unquestionably, was a human being.

Ergaster was an obligate biped that had abandoned adaptations to tree climbing, and who could probably climb no better than we can. Ergaster would have to make its living on terra firma. So, why had it forsaken the hybrid terrestrial/ arboreal lifestyle that had served Australopithecus so well for so long? Once again, it was climate change.

This was now the Pleistocene Epoch (2.6 MYA to the present). In Africa and southern Asia, the climate had grown even dryer and cooler. Grasslands expanded and the forests diminished further. With forest food becoming rare, Ergaster would have to rely solely on the resources of the savanna. Luckily, there was ample nutrition out there, mostly on the hoof. Accessing it, however, would be another matter altogether.

Australopithecus, as you will recall, was omnivorous, capturing and devouring small animals and perhaps occasionally power scavenging the fresh kills of actual predators. Their diet, however, was mostly plant food, and, thus, they remained rather small of body and brain throughout their existence. Ergaster would need a more consistent supply of animal products to sustain itself, and there was only one way to do that; hunting.

But how? Savanna quadrupeds were far faster than they, and Ergaster had no bows and arrows or other long-range weapons at its disposal. There was, however, one thing that Ergaster could do that most grassland mammals could not and that was long distance running.

Modern humans have inherited from Ergaster a number of traits that are useful only for long distance running, including:

Long, springy tendons in our legs and feet that act as elastic bands that store and release energy;

Enlarged gluteus maximus muscles that hold our bodies upright while on the move;

The ability to rotate our shoulders independently from our heads for better balance while running;

A nuchal ligament at the back of our head to counter the head's tendency to pitch forward when running;

Large lumbar vertebrae to enhance shock absorption.

Such a major overhaul of the human body plan can only mean that long distance running once offered a significant survival advantage.

Ergaster also developed a number of adaptations to deal with the enormous amount of heat generated by running, particularly when running beneath a tropical sun. Ergaster lost its body hair early on, retaining, like us, only a patch on its head to protect its skull from the sun. It developed sweat glands, which were concentrated in the scalp to cool the brain and on the front surfaces of the torso and limbs over which air could pass as the body moves forward.

Other animals of the savanna, then as well as now, dissipated heat through panting. They could not, however, pant while galloping, requiring regular rest stops to cool off. Hunting then simply became a matter of spotting a suitable animal on the open plain under the midday sun, giving chase, tracking it if it got out of sight and flushing it when it tried to rest. Under such harassment, most animals would succumb to heat exhaustion within 10 to 15 kilometers, becoming easy marks for Ergaster's clubs and spears. This form of persistence hunting is still practiced today among some aboriginal peoples, and with great success.

Apes, including Australopithecus, have a large gut, or lower GI tract, which is a significant impediment to long distance running. The gut includes the small and large intestines. Higher quality foods, such as animal fat and protein and soft fruit, are digested in the small intestines, whereas lower quality foods, such as leaves, stems and fibrous fruit, are digested in the large intestines. Apes derive much of their metabolic energy through fermentation of lower quality plant material in their hindgut. As a result, the large intestines are truly large, comprising the majority of their gut, being about two to three times the length of their small intestines. As human populations improved their diet through animal products, the ratio between large and small intestines was reversed, with human small intestines being three to four times the length of their large intestines. This resulted in humans evolving a significantly diminished gut, making long distance running possible.

The evolution of the ability to run and a reduced gut seems to have been mutually reinforcing, where long distance running enabled Ergaster to hunt, which provided better nutrition, which allowed the size of the gut to diminish, which enhanced the ability to run, which resulted in better hunting and so forth. Through hunting, the diet of the human genus, as well as the lifestyle that supported it, was forever changed. Humans had lost the capacity to live on the fibrous, vegetarian diet of the apes and even Australopithecus and had become, to a large degree, carnivores.

The higher quality food now required by humans was more nutritionally dense and provided significantly more calories. This allowed both their bodies and brains to grow. Already, by 1.6 MYA, Turkana Boy and his kind were substantially bigger than Australopithecus, and their brains nearly fifty percent larger. And this was just the beginning.

The importance of dexterous hands to human development cannot be overstated. Without the capacity to skillfully manipulate objects in our environment, our big brains would be pointless. Indeed, they have evolved in tandem with our hands.

Ape hands are long with long, curved fingers that are well adapted to climbing and swinging through the trees.

Modern human hands, on the other hand, are smaller, and our fingers are straight and significantly shorter than those of the apes. Also, our thumbs are opposable and longer relative to our fingers, allowing us to firmly grasp objects of many different shapes and sizes. Our hands are a miracle of biological engineering, consisting of twenty-nine bones, the same number of joints, thirty-five separate muscles, a hundred tendon and a vast network of nerves and arteries. Together, they allow our fingers to move independently and operate with fine control.

As we have seen, the movement toward more dexterous hands began with Australopithecus. Yet, they remained tethered to their terrestrial/arboreal hybrid lifestyle, limiting the development of their hands. Ergaster, however, had forsaken the trees. Freed of the need of adaptations for climbing, Ergaster's hands quickly evolved the ability to make and use more complex tools. The feedback between more dexterous hands, leading to more sophisticated tool use, leading to even more dexterous hands and so forth contributed significantly to its acquisition of greater cognitive ability and brain size.

Despite their increased brain size and evolving hands, Ergaster continued for a very long time to use the same old Oldowan, or Mode I, tools that had been in use by Australopithecus for a million years. By 1.5 MYA, however, a new technology had replaced the sharp flakes and cobles of old. The new tools were the so-called hand-axes. These were rock cobles flaked over on both sides to produce a sharp edge around the entire periphery. Typically, they were eight to nine inches long and teardrop-shaped, narrowing from a broad base on one end to a round point on the other. They were employed for diverse purposes, including butchering animals, scraping hides and chopping wood. Hand-axes, considered Mode or

Level 2 technology, would remain state of the art for more than a million years.

Hand-axes, though seemingly simple to us, were clearly beyond the abilities of apes to produce. Rather than just producing sharp flakes, the maker had to choose the right size and kind of rock, free of flaws, then trim the rock in such a way as to produce two cutting edges that converged to a pointed tip. Thus, to make the hand-axe, the maker had to envision the finished product before he even started. This required cognitive capabilities that were far beyond any that had so far existed.

That it took so long for Ergaster to develop this improved technology should not be surprising. First, tools have to be invented by someone, which, in a primitive world, would not have been an easy thing to do. Then, others have to learn how to make the new tool, which, at a time without modern language, would require the student to actually observe the maker making the tool, probably several times, followed by the student's making a number of unsuccessful attempts before finally producing the finished product. At a time of limited contact between groups, the spread of the new technology was sure to have been painfully slow.

So, where did Ergaster come from? Certainly from Australopithecus, but there is no consensus as to which one. The problem is that Ergaster's body plan is radically different from that of Australopithecus, but there are no intermediaries. Ergaster seems to have popped up out of thin air. But then, that's the way evolution works sometimes.

As we have seen, Darwin believed that species changed gradually and incrementally. One species simply morphed into another by the gradual accumulation of small changes over time. That is not, however, what the fossil record shows. Species tend to arise rapidly, followed by a long period of stasis.

A reason for this is the way genes work. There is not a single coding gene for each physical characteristic. Otherwise, there would be a vast number of different coding genes, which there are not. Instead, most coding genes are involved in determining a number of different characteristics in conjunction with other coding genes. Moreover, each physical characteristic requires the coding genes to be switched on at a specific time and for a specific duration by other genes referred to as switching genes. Then, there are regulatory genes that govern the vigor with which the coding genes are expressed. A small mutation that causes differences in gene timing and expression can result in huge differences in the individual's physical features. Typically, such a radical change in morphology will render the individual defective and unable to survive. Occasionally, however, the new features will open up a new and better way of exploiting the environment, quickly giving rise to an entirely new species. That may indeed be what occurred with Ergaster.

In the early 1990s, in the ruins of the medieval town of Dmanisi in the Republic of Georgia between the Black and Caspian Seas, specimens of Homo Ergaster were discovered in sediment dated to 1.8 MYA. That was only about a hundred thousand years after the species had evolved. As would be expected, the individuals were very early forms of the species, having smaller bodies and smaller brains (600 cc to 775 cc) than Turkana Boy and his kin, who lived 1.6 MYA and whose brain volume averaged about 850 cc. The tools found with them were simple flakes and did not include hand-axes, which had yet to be invented. How, then, did such early forms of Ergaster, with still small brains and only simple tools, travel the 2300 miles separating Ethiopia from Georgia? The short answer is they did not. With the general drying of the climate, the grasslands of Africa, and the grazers who live on them, spread into Eurasia. Ergaster, then, simply followed along with the expanding Savanna. 2300 miles may seem like a very long distance to travel, but over a period of 100,000 years that works out to roughly 120 feet per year, which is about the roundtrip distance from my easy chair to my mailbox on the street. Though the species as a whole covered a vast distance in that time, it is unlikely that any single individual would have had the sense that he had ever gone anywhere.

One of the skulls found at Dmanisi was that of an aged individual who had during life lost all of his teeth but one. Since you can't chew a whole lot with only one tooth, he faced a substantial risk of starvation. Clearly, the old one could not have survived without the constant, long term assistance of his family and the other members of his group. This is the first evidence of altruism in the fossil record, though many similar examples would follow.

By contrast, researchers have conducted a series of tests on groups of chimpanzees to determine whether they were capable of the kind of empathy expressed by the Dmanisis group for the old, one-toothed man. In various ways, the chimps were given the option of obtaining a food reward for itself and for its companion or only for itself, the task and reward in either case being the same. The chimps chose pretty much at random, indicating no real regard for the needs of its fellows. Altruism, then, appears to be an exclusively human trait, a trait which developed at the very beginning of our genus.

Ergaster did not stop in central Asia but continued eastward, following the expanding grasslands. By 1.5 MYA, they had reached eastern Asia, where they flourished until as recently as 100,000 years ago. Because of their early and long separation from the African branch of the species, the eastern group shared a number of regional variations that differentiate them from the African group, but they all shared the same basic body plan. Despite their long and successful tenure in east Asia, they eventually went extinct, leaving no descendants. Consequently, they are not a part of our story, which will continue in Africa, Europe and western Asia.

But first, a word about names. The earliest fossils of Homo Ergaster in Africa were not discovered until 1975. The first fossil of the species, however, was discovered by the Dutch physician, Eugene Dubois, on the island of Java in Indonesia in 1891. The fossils were colloquially referred to as Java Man but were later given the name of Homo Erectus (Upright Man). Since the name was applied to other fossils of the species subsequently found in China and other parts of the east, the named Homo Erectus had been in use for a long time by 1975. When it was generally agreed that Ergaster and Erectus were the same or very closely related species, many paleontologists, using the more established name, referred to the African group as the African Homo Erectus of just Homo Erectus. Just bear in mind that whether called Ergaster, African Homo Erectus or just Homo Erectus all refer to the same species. A rose is a rose is a rose.

Ergaster also spread throughout much of Africa. The fossil record of these times, however, is scant and muddled and, consequently, difficult to interpret. We do know that a new species of genus Homo appeared about 600,000 years ago. The first fossil of this new species was a lower jawbone found in 1907 near the German town of Heidelberg. It was thus given the name Homo Heidelbergensis. Since then, fossils of Heidelbergensis have been found throughout much of Africa and Europe.

Heidelbergensis retained much of the primitive look of Ergaster, with a large, flat, forward-projecting face, a massive chinless lower jaw, a heavy brow ridge and a short, sloping forehead. Its brain, however, was large, weighing about 1250 cc, which is approaching the size of modern humans at 1350 cc. The striking increase in cranial capacity appears rather suddenly in the fossil record and may have been the result of a punctuated event, as had occurred with Ergaster. The increase in brain size was likely related to the worsening climate swings that Europe and, to a lesser extent, Africa were experiencing during the middle Pleistocene. Survival might have depended upon more creative responses to the challenging new conditions.

Not only its brain but Heidelbergensis' body was also growing larger. Post cranial bones are rare, but a 500,000-year-old shin bone was found in the village of Boxgrove in West Sussex. It is one of the most massive leg bones of an early human ever found. The possessor, believed to have been a man of about forty at his death, would have stood nearly six feet tall and weighed more than 200 pounds.

Heidelbergensis was an accomplished hunter, as is attested by the numerous butchered animal bones found at their sites. Unlike Australopithecus and Ergaster, Heidelbergensis actively hunted its prey. Any doubt as to their hunting prowess was removed with the 1995 discovery of eight wooden spears in Schoningen, Germany. The spears had somehow been preserved in a lake bottom for the 400,000 years since their manufacture. Each was about six and half feet long, scraped smooth with stones and finely sharpened. The spears were weighted toward the front, suggesting that they had been used as javelins, rather than as simple thrusting spears. The bones of ten butchered horses were also found at the site. That Heidelbergensis regularly hunted big game is also supported by the discovery at the Boxgrove site of a 500,000-year-old fossilized rhinoceros shoulder blade with a projectile wound, thought to have been made by such a spear.

The Schoningen site also yielded the first composite tools, that is, tools made from more than one component. The importance of such tools can be grasped by imagining trying to chop down a tree with an axe head but no handle. Sharp edged flints were found next to foot-long, grooved wooden tools to which the flints had apparently once been hafted.

Besides advanced weapons and composite tools, Heidelbergensis developed an entirely new tool technology, referred to as prepared core tools. This Mode 3 technology involves shaping a stone core to a pre-planned design, then detaching a flake of a desired size and form with a single blow from a hammer made of a soft material, such as bone. The flakes could then be fashioned into distinct types of tools, each with its own function, such as cutting, scraping or piercing. Since the making of such tools requires the maker to plan several stages ahead, the technology represents a new level of cognitive complexity.

Heidelbergensis also constructed the first artificial shelters. At Terra Amata in southern France, archeologists have found evidence of a number of large huts dating back 400,000 years. These oval shaped huts were built by embedding saplings into the ground then bending the tops toward the center where they were tied off. A ring of stones was place around the hut to brace the saplings.

Near the entrance of some of the huts, a hearth had been scooped out of the ground and lined with stones. Remnants of charcoal lined the bottom of the hearths. This is the oldest definitive evidence of fire use. There is some evidence of fire use by early humans as far back as a million years ago, though much of it is ambiguous. But even if the claims of suspected fire use were all true, such use would have been infrequent and sporadic, indicating the occasional use of captured wildfire rather than the ability to produce fire. After 400,000 years ago, on the other hand, evidence of fire use is regularly found at sites of human occupation.

Not only did the Heidelbergensis make and use fire for warmth and light, they appear to have been the first of the early humans to regularly use fire for cooking. The copious charred animal bones often found associated with the hearths make it clear that they did indeed cook their food. Cooking with heat breaks down the collagen fibers that make meat tough and soften the cell walls of plant foods, making the nutrients within more accessible. Cooked food is far more digestible than raw food, greatly increasing the number of calories derived from the food. The increase in the amount of meat in its diet through regular and successful hunting and the ability to cook its food enabled Heidelbergensis to grow both its body and its brain.

Heidelbergensis walked the Earth from 600,00 to 200,000 years ago. Though not the radical, groundbreaking species that Ergaster had been, it did much to improve the human condition, engaging in true hunting and crafting the weapons necessary to do so, devising composite tools and the prepared core technology, building shelters and learning to make and use fire for warmth and cooking. Perhaps, though, its greatest contribution to the genus was begetting two new species, Homo Neanderthal in Europe and Homo Sapiens in Africa, which is where we will turn next.

6. Sapiens

Origins. The fossil evidence indicates that Homo Sapiens evolved in Africa from the African version of Homo Heidelbergensis. Anatomically modern Homo Sapiens first appeared about 200,000 years ago. Whether modern morphology appeared suddenly or whether it developed slowly between 400,000 and 200,000 years ago is a topic of fierce debate among paleoanthropologists.

The skulls of anatomically modern humans differ markedly from those of earlier members of our genus. Our skulls have a globular braincase with the face and eyes tucked underneath, the forehead is steep and the chin prominent. In earlier humans, like Ergaster and Heidelbergensis, the braincase was long, low and oval shaped, the face jutted forward and the forehead was flat and receding.

By 400,000 years ago, there were a number of large-brained early human populations spread throughout Africa, some with brains as large or even larger than our own. They did not look like us, however, retaining the ancestral look shared by all previous members of the genus Homo. Most, though, had at least some modern characteristics.

A case in point are cranial and post cranial fossils of early humans that lived 300,000 years ago found in the Jebel Irhoud cave near Marrakesh, Morocco. Their brains are estimated to have been between 1300 and 1480 cc, which is well within the modern range. Their braincases, however, were elongated, with thin brow ridges and very large teeth, characteristic of archaic species of Homo. On the other hand, their faces were tucked under the skull rather than jutting forward, as do the faces of modern humans.

This mix of modern and archaic features found in the Jebel Irhoud fossils are also found in other populations all across Africa during this period. This has led many paleontologists to conclude that Homo Sapiens did not emerge at a single place and time in a single population. Instead, numerous isolated populations, each possessing some modern characteristics, eventually encountered and interbred with each other. Over time, a single population possessing all of the features of modern humans emerged from the mix.

Other paleontologists strenuously disagree with this "accretion" or "Pan African" model, pointing out that there is simply no basis for identifying as "archaic" or "early" Homo Sapiens a grab bag of human populations from all over Africa who, although large brained, do not share many of the most basic aspects of our distinctive anatomy. Not only are these archaic populations not Homo Sapiens, they are likely only distantly related, long extinct, historical dead ends. For these scientists, the qualitative difference between Sapiens and the rest of the genus was as profound as it was between Ergaster and Australopithecus. The emergence of such an entirely new species, they argue, could only have occurred in a small population over a short interval. The matter of our origins, it seems, remains contentious.

However we got here we did get here by about 200,000 years ago. In 1967, Richard Leakey and his team working in the Omo River basin in southwest Ethiopia uncovered two skulls, one fragmentary with some postcranial remains and the other nearly intact. The skulls have since been dated to 195,000 years ago. They are referred to sensibly enough as Omo I and Omo II. Omo I, as reconstructed, has a globular braincase, a steep forehead, a prominent chin, small brow ridges and teeth of modern size and shape. Though quite robust, it is clearly the skull of a Homo Sapiens, the oldest ever discovered. Omo II had a large cranial capacity of 1435 cc, but its brain case was long and its forehead receding, among other primitive features, and was not a Homo Sapiens. The two skulls, though found in the same geologic stratum, were probably not exact contemporaries, likely living several centuries apart. Omo II was apparently a late surviving archaic form.

In 1997, three well preserved skulls of two adults and one child were found in Herto Bouri in Ethiopia. The remains were dated to between 154 and 160 KYA. The braincases were globular, with the adults having a cranial capacity of 1450 cc, and the facial features were more or less modern. Though classified as a Homo Sapiens, the Herto people retained some archaic features, such as a longer braincase and more robust features, and so are called by some "early" Homo Sapiens.

Homo Sapiens remains have also been found at the Jebel Qafzeh site in Israel. The fossils date to 115,000 years ago, which is the latter part of the warm Eemian interglacial period, when the Sahara was wet and the Levant accessible to human populations living in northern Africa. Also found at the site and at nearby Mugharet es-Skhul were fossils of other individuals who had very large brains (with cranial capacities of up to 1590 cc) but who retained a number of archaic features and do not appear to have been of our species.

Despite disagreements about how our species originated, paleontologists are in general agreement that anatomically modern humans had spread throughout much of Africa by about 150,000 years ago. Behavioral modernity, however, would take a while longer to develop.

Cognition. Unique among all animals, even our closest relatives, modern humans have the ability to think symbolically. Symbolic thinking is the ability to use symbols, concepts and abstract relationships to analyze the world around us. It makes language, literature, numeracy, science and artistic expression possible, and it is the basis for modern human behavior.

Anatomically modern humans emerged some 200,000 years ago, but they were not behaviorally modern. For a very long time, they continued to live as their ancestors had lived and as other large-brained contemporary early humans would continue to live, using the same tools and engaging in the same lifestyle.

After about 65,000 years ago, however, the archeological record is replete with artifacts that unquestionably reflect the workings of cognizant minds. Cave art depicting animals and spiritual beings began to be produced in Europe, Australia and elsewhere 40,000 years ago. Portable artwork, such as carved ivory statuettes and figurines were widespread by 28,000 years ago. The first musical instruments appear 32,000 years ago. Evidence of the use of fitted clothing, such as sewing needles, and adornments, such as buttons and beads sewn into the clothing, appear 30,000 years ago. Burials, which were infrequent and simple before 40,000 years ago, became more abundant and elaborate after that date. The production of small blades called microliths (Mode 5 technology) were regularly being hafted onto arrows, barbed spears and sickles by 40,000 years ago.

The seemingly sudden appearance of behavioral modernity at this time has been termed by the anthropologist, historian and all-around thinker, Jared Diamond, as the Great Leap Forward. Proponents of this view postulate that such sudden advances could only have come about as a result of a genetic mutation that enabled Homo Sapiens to think symbolically. The specific mutation, though, has never been identified. Moreover, there are a number of finds indicating that the great leap forward may not have been as sudden or great as had been previously supposed.

At Pinnacle Point in South Africa, for instance, there is evidence that, as early as 165,000 years ago, some early Homo Sapiens were using red ochre as a pigment for body painting. Also, they were hafting microliths onto bone or wooden handles and to the tips of spears. This Mode 5 technology disappears after Pinnacle Point for more than 100,000 years.

At the Katanda site in the Congo, a number of bone artifacts dating from 90,000 years ago have been found, including barbed points that might have been used for river fishing. Bone tools, though offering a number of technological advantages over stone tools, are labor intensive and time consuming to make and, therefore, are considered signs of modern human behavior.

At the Blombos Cave in South Africa, numerous artifacts dating to 73,000 years ago have been found. Perhaps most notable are two ochre plaques, each several inches long and engraved with a crosshatch pattern. These may be the world's oldest works of art and are the earliest objects we know of that are undoubtedly symbolic. Also found were a number of beads made from perforated shells of the marine snail, Nassarius. Tool use was also advanced. The Blombos people used compound adhesives to attach spear points to shafts. They also used the pressure flaking method to finish stone blades. This technique requires a stone blade to be heat treated to improve its flaking qualities. Then, pressure is exerted near the edge of a blade using a pointed stick or bone to detach small flakes from both sides of the blade. This complex procedure was once thought to have been in use no earlier than 20,000 years ago.

Finally, at another site in South Africa, the Diepkloof Rock Shelter, pieces of twenty-five ostrich eggshells, each engraved with standardized geometric patterns, have been recovered from a 65,000-year-old layer of occupation. The shells had, apparently, been used as water flasks, as they are still used today by peoples of the Kalahari.

These examples demonstrate that the ability to think symbolically has likely been with our species from the very beginning. If so, the anatomical changes that reshaped our skulls and the neurological changes in the brain necessary for symbolic thought in Homo Sapiens must have occurred at the same time. There was, then, no "smartness mutation" that came along later.

The braincases (and brains) of other species of the genus Homo grew over time, but the shape did not change. Consequently, they were just larger versions of what had come before. The globular braincase of Homo Sapiens, by contrast, reflects internal changes to the actual configuration of the brain. It is this reorganization and rewiring of the brain that may have led to a new level in complexity of function.

As noted above, there is very little in the fossil record between 200,000 to 100,000 years ago showing that Homo Sapiens had acquired a higher degree of cognitive complexity. By about 100,000 years ago, however, there is clear evidence of a budding new level of cognition, and, by 65,000 years ago, our species was exhibiting essentially modern behavior. If Homo Sapiens were capable of modern behavior from the outset, why did it take so long for such behavior to manifest itself? The reason may be that the neural restructuring that resulted from the anatomical changes in our brains enabled but did not cause our brains to make the complex associations required by symbolic thought. Homo Sapiens would still have to learn to use these new functions, and that would take time. In the meantime, they simply continued to employ the old style of brain function.

The means by which this new cognitive function would eventually be fully exploited was, in all likelihood, language. Syntactical language is the ultimate symbolic activity. It can generate an infinite number of concepts from a finite number of elements, making thought as we know it possible. It is uniquely human and is hardwired into our brains. Unlike reading, which must be taught, young children pick up language almost effortlessly and without instruction. Chimps can be taught a few signs, but they cannot string the signs together or combine them to express novel concepts. Syntactical language is far beyond their capacity, as it almost certainly was with all of the other species of our genus. Although the ability to use language appears to have been innate in early Homo Sapiens, language itself still had to be invented.

When this occurred is difficult to say, since language leaves no trace in the fossil record. Moreover, because multiple regions of the brain are involved in language, attempts to identify the specific changes in the anatomy of the brain that made language possible have enjoyed little success. Language does, however, serve to facilitate the development and spread the use of other forms of symbolic thought. Thus, tangible evidence of symbolic cognition in the archeological record, such as works of visual art and advanced tools, infers the presence of language. On that basis, it is reasonable to conclude that language was widespread by 65,000 years ago.

The transition to a symbolic, linguistic means of processing and communicating information was a qualitative leap in the cognitive state of Homo Sapiens. With it, the species had become behaviorally modern and was ready to spread across the planet.

Diaspora. Because the archeological record is sparse, discerning human population movements in the distant past is extremely difficult. There is, however, another way. Since all people are descended from an ancestral population, geneticists can trace ancient movements by analyzing the DNA of living people and working backwards.

As the result of sexual reproduction, the offspring acquire from their parents genetic material that has been shuffled or recombined, making it difficult to trace evolutionary relationships within a species. The X and Y chromosomes, however, do not exchange segments of DNA. This is to ensure that the gene in the Y chromosome that makes one male never gets spliced into the X chromosome. Thus, the Y chromosome gets passed down unchanged from father to son, generation after generation.

Mitochondria, as we have seen, are once captured bacteria living within eukaryotic cells that generate energy for use by the cell. Mitochondria are located in the main body of the cell outside of the nucleus. When, during reproduction, the sperm meets the egg, the sperm's genetic material is fused with the genetic material in the nucleus of the egg. The remainder of the sperm is discarded. Consequently, only the egg's mitochondria survive. That means that the mother's mitochondria are handed down to her children unchanged through the generations. As relic bacteria, mitochondria have their own simple loop of DNA, which does not mix with the cell's nuclear DNA and so remains essentially unchanged.

Although the Y chromosomes of males in subsequent generations are copies of the ancestral Y chromosome and the mitochondrial DNA of all subsequent females are copies of the original mitochondrial DNA, they are not identical copies. Over time, small mutations in the DNA occur at a steady rate, which are then passed down to subsequent generations. Thus, a mutation in the mother's mitochondrial DNA will be passed down to her daughters, then to their daughters and so on. Likewise, a mutation in the Y chromosome will be passed down through the male line to all males in subsequent generations. Anyone carrying one of these specific mutations can be identified as being related to others with that same mutation. By examining the DNA of living persons, a genealogy of humankind based on the Y chromosome and a separate genealogy for mitochondrial DNA can be mapped. Because, over history, most people have lived their lives in the same geographical area where they were born, these genealogies can be used to track population movements over time.

Genetic studies show that, until modern times, a few men in Sub-Saharan Africa and all men in the rest of the world carried a Y Chromosome mutation known as M168. Studies also show that there are three main branches of mitochondrial DNA, referred to as L1, L2 and L3. L1 and L2 are found only in Sub-Saharan Africa, while everyone outside of Africa carried three sublineages of L3 known as M, N and R.

The Y chromosome and mitochondrial data indicate that, about 65,000 years ago, a small group of Homo Sapiens living in east Africa, perhaps as few as 150 individuals, left the continent and over time populated the rest of the world. Their route out of Africa and dispersal around the globe, in broad brush, occurred as follows.

The Bab el-Mandeb Strait is a narrow passage at the southern end of the Red Sea separating Africa from the Arabian Peninsula. Today the crossing is eleven miles. 65,000 years ago during a particularly nasty period of the last ice age sea levels were more than 200 feet lower and the strait less than seven miles across. As the coastal people were almost certainly fisherfolk skilled in the use of boats, the crossing would not have been an especially difficult feat.

Once on the other side of the strait, the travelers would have found life there not so much different from their lives back in Africa. Fishing and beachcombing would have amply provided for their needs, just as they had always done. As populations grew and resources became strained, groups could simply break off and move to virgin territory farther down the coast. In this way, over the generations they moved along the southern coast of western Asia to India then around the subcontinent into Southeast Asia.

Because sea levels were so much lower than they are now, today's islands of Sumatra, Java and Borneo, along with the Malay peninsula, formed a large landmass extending south of present-day Southeast Asia, which has been named Sundaland. At that time, Australia was joined to New Guinea in the north and Tasmania in the south to form a continent known as Sahul, which lay directly south of Sundaland. A deep-water channel dotted with islands separated Sundaland from Sahul. To reach Sahul, however, required as many as seventeen separate island-hopping crossings, some as long as 44 miles. Doing so would have necessitated efficient nautical technology, which the colonizers must have had, since they reached Australia by 50,000 to 55,000 years ago.

The long trek from east Africa to Australia is a good story. There is, however, no archaeological evidence to support it. In fact, the oldest archeological site of modern humans outside of Africa is in Australia. The lack of physical evidence of the long journey should not be surprising, though. With the rise in sea levels since that time, the beaches comprising the coastal route are now miles offshore, and any archaeological evidence of the migration was swept away thousands of years ago. Although there are no artifacts or fossils to prove that the founding population followed the beachcombing route to Australia, there is another kind of evidence of their journey, human evidence.

Scattered along the beachcombing route are isolated populations of people anthropologists refer to as "negritos" (little black people) (their word, not mine). As the name implies, they are dark skinned with tightly curled hair and look all the world like indigenous Africans. They are found in the Andaman Islands in the Bay of Bengal, the Malay peninsula, Thailand and the Philippines, all of which are thousands of miles from Africa. It was long a mystery how isolated populations of African peoples could be found so far from their homeland. Genetic studies show that these peoples settled those locations 60,000 years ago, and there they remained isolated from mainland populations until modern times. In other words, they are relic populations of the original migration out of Africa.

It would seem that the beachcombers in Southeast Asia would have continued eastward around the southern coast of Sundaland, then northward up the eastern coast of China to Japan and the Kamchatka peninsula. That is apparently not what happened. In 2011, a genetic study of Aboriginal Australians found that Han Chinese (the dominant ethnic group in China) are more closely related to Europeans than to the Aboriginals. The study shows that the split between the Han and the Aboriginals occurred around 62,000 years ago, while the Han and Europeans did not split until 10,000 to 15,000 years later. It appears, then, that Australia was settled by a first wave of beachcombing migrants and East Asia by a second wave thousands of years later. The second wave followed the coastal route to Southeast Asia but then continued around Sundaland rather than heading south into Australia. All three groups, however, are equally distantly related to Africans, confirming that both waves were descended from the original founding population.

Not all of the migrants followed the beachcombing route. Many moved inland along river systems and rapidly settled the interior of India and Southeast Asia. Indeed, by 45,000 years ago, the population there exceeded the human population of Africa.

Some migrants in India later moved north into the Levant and from there to Europe, entering southeastern Europe by 46,000 years ago. Some groups moved rapidly across Europe from east to west via a northern rout along the Danube River and a southern route along the Mediterranean coast. The two routes converged in southern France 41,000 years ago then moved south into Spain. Other groups moved north into the

East European Plain, reaching southern Russia around 42,000 years ago. Some of these then moved as far north as the Arctic Circle.

The Eurasian Steppe stretches from the Hungarian Plain to Mongolia. Its climate was arid, cold and windy, but, 40,000 to 50,000 years ago, it was grazed by vast herds of mammoths, wooly rhinoceros, wild horses, reindeer, musk oxen, camels and antelopes. Despite the hostile conditions, hunting opportunities lured people northward. Genetic evidence indicates that, by 49,000 years ago, groups were moving onto the steppe from India, Southeast Asia and China. For a long time, the settlers remained in the relatively milder climates below 55 degrees North Latitude. By 30,000 years ago, however, they had learned to cope with the harsh conditions and began to settle lands further north, some even moving above the Arctic Circle.

Today, Siberia and Alaska are separated by the Bering Strait, a distance of only 18 miles. From 25,000 years ago until 10,000 years ago, however, sea levels were so low that the strait and parts of the Chukchi Sea, to the north, and the Bering Sea, to the south, were dry land, forming a land bridge between the two continents. This region, referred to as Beringia, was actually a bit more than a bridge, stretching as it did some 3,000 miles from the Lena River in Siberia to the Makenzie River in the Yukon and a thousand miles from north to south. Though quite cold, Beringia remained ice free because of the extremely arid conditions of the region. The land consisted of steppe-tundra and was home to huge numbers of steppe animals, great and small. The first fossil evidence of human presence in Beringia dates only to about 14,000 years ago, but DNA studies indicate that the region was occupied much earlier than that.

Expansion into the Americas was blocked by the Laurentide Ice Sheet, which covered most of Canada and a large portion of the northern United States as far south as Pennsylvania, and the Cordilleran Ice Sheet, which extended from western Alaska to northern Washington, Idaho and Montana. At the end of the Last Global Maximum, the ice began to retreat. By 15,000 years ago, a route along the Pacific became deglaciated and open to human habitation. Additionally, a corridor between the two ice sheets opened up between 14,000 and 13,500 years ago, providing another route for occupation of the Americas.

It had long been assumed that native Americans were descended from today's indigenous peoples of eastern Siberia. Genetic studies reveal, however, that they are actually much more closely related to peoples of the eastern part of Central Asia. Thus, it is now believed that the ancestors of Native Americans began migrating from Central Asia to the northeast 46,000 years ago, moving into Beringia 30,000 years ago. During the Last Global Maximum, the people of Beringia became isolated, with the way into North America blocked by the ice sheets and eastern Siberia having become a frigid, desert wasteland. As the climate warmed around 16,000 years ago, they expanded into Alaska and then along the Pacific route into the Americas. From there they moved rapidly through North America, reaching South America by 14,000 years ago and all the way to Tierra Del Fuego at the continent's southern tip by 10,000 years ago. With that, Homo Sapiens had populated the entirety of the Earth's habitable continental landmasses, a feat not previously accomplished by any other large mammal species.

Revolution. By 65,000 years ago, Homo Sapiens were behaviorally modern and fully capable of symbolic thought. They remained, however, nomadic hunter-gatherers, following the game, making seasonal treks to preferred hunting grounds and generally living as they had always lived and as our genus had lived for hundreds of thousands of years.

Such a lifestyle, I imagine, had its charms. Hunter-gatherer societies typically have access to a great diversity of wild foods, providing them a nutritionally balanced diet and requiring only a few hours of hunting and gathering a day to meet their dietary needs. Crop failure is not a danger to these societies. Kalahari Bushmen, for instance, eat some seventy-five varieties of wild plants, making a potato blight style famine almost inconceivable.

Hunter-Gatherers also have far more leisure time on their hands than do settled folk. Once they have eaten, there really isn't much anything else for them to do...no barns to paint or fences to mend or ditches to clear or chickens to feed. Nothing to do but kick back by the campfire and shoot the bull.

These societies would have been highly egalitarian, with no rich guys lording it over the less fortunate. Back then, you only owned as much property as you could carry, and, since you were always on the move, that would not have been much ... a spear, perhaps, and a knife and maybe a trinket or two, same as everybody else.

They would have been quite healthy as well. Isolation would have made contagious diseases rare, and natural selection would have swept away genetic proclivities for poor eyesight, bad teeth, heart murmurs and the like. Moreover, many of today's maladies, like the influenza, tuberculosis, malaria, bubonic plague, measles and cholera, evolved from diseases infecting farm animals, of which there were none at the time. Moreover, the aged would not have fared well in such conditions, so the population would have been very youthful.

Healthy, young, well fed, leisurely and free. It all sounds idyllic, but there was trouble in paradise.

In primitive societies, women kept the family functioning. They raised the children, cooked the food and gathered the berries, nuts, tubers and roots that were the staple of the clan's diet. When the men occasionally brought home game, the women cleaned and prepared the game as well. Men in such societies didn't have a whole lot of daily responsibilities. Besides hunting, the men's primary activity was fighting. Archaeological evidence from the Upper Paleolithic period (50,000 to 10,000 years ago) and anthropological studies of primitive peoples, including the !Kung San bushmen of southern Africa, Eskimos, Australian Aborigines, the Yanomamo Indians of Brazil and the pig and yam societies of New Guinea, show that warfare in pre-state societies was very frequent. It appears that more than half of such societies were at war continuously, and ninety percent fought more than once a year. Pitched battles were rare, most preferring ambush and dawn raids. The objective most often was to surprise and kill one or more of the enemy, then flee home without being discovered. Rape and capture of the enemy's women were expected benefits but not the motivating factor for a raid. Casualties were generally light, but, because fighting was so prevalent, they mounted over time. It has been estimated that tribal societies lost on average 0.5% of their population in combat every year. That doesn't sound like much, but it actually is. Around a hundred million people died in combat during the bloody Twentieth Century. At 0.5% of the world's population annually, that number would have been two billion.

The biological basis for such aggressive behavior was the prestige and consequent reproductive advantage gained by those who had killed an enemy. Among the Yanomamo, a man who had killed an enemy was found to have on average

2.5 times as many wives and three times as many children as a man who had not. A propensity for violence, martial values and hostility to outsiders, then, were bred into our species. Among the Iroquois, a young brave prepared himself for the day he might be captured and dragged back to the camp of an enemy, where he would be feasted and encouraged to boast of his exploits, after which he would be tortured mercilessly for hours, his job being to bear the blows and burns and cuts stoically, to hurl insults at his tormentors and then to die magnificently.

Such were our forebearers, gifted with symbolic thought, language and art but living in ignorance, superstition and violence. For some 50,000 years, they roamed, hunted and fought, building nothing, creating little and leading lives that were essentially the same as those of the generations that preceded them. This long period of stasis, however, would eventually come to an end. The catalyst for a new lifestyle was once again climate change.

The most recent ice age (115,000 – 11,700 years ago) saved its worst for last. Between 22,000 and 19,000 years ago, in the period referred to as the Last Glacial Maximum, the ice sheets expanded to their greatest extent. Much of North America, northern Europe and northern Asia were covered in as much as two and half miles of ice, and sea levels dropped by 400 feet. Thereafter, the Earth remained cold until the onset of the sudden warming event known as the Bolling-Allerod Interstadial 14,700 years ago. This warm period would last for nearly 2,000 years. In southwest Asia, the climate became warmer, wetter and more seasonal, permitting oak woodlands and grasslands to spread rapidly. In this new setting, a different form of human society began to emerge, one based not on the nomadic hunter-gatherer lifestyle but on sedentism, that is, settling down in one spot.

Archaeologists consider the advent of sedentism a revolution as significant as the emergence of behaviorally modern humans from their anatomically modern ancestors 65,000 years ago. Sedentism was a prerequisite to civilization, and it was only in civilized society that the full human intellectual capacity could be realized. Thus, the transition from foraging to settlement was crucial to human cognitive development. It wasn't easy, though.

There were advantages to living in settled communities, the most significant of which was allowing people for the first time to gather and store food surpluses. Additional food meant more mouths could be fed, resulting in a rapid rise in population. The villagers would have soon encroached the territories of the hunter-gatherers, but their greater numbers would have deterred attacks by the small nomadic bands.

Community life, however, required significant adjustments on the part of the residents. They now had to deal with intellectual concepts that were alien to forager societies, such as private property, the general welfare and the common good, and cope with rules, regulations, limitations and obligations that had not existed before. Moreover, life in settled societies required significant behavioral modifications, not the least of which was just being able to live in close proximity with others who were not related. The forager life had bred into the human psyche a deep hostility to strangers. It may be that that some villagers were more genetically predisposed than others to living nearby and cooperating with people not their kin. Those who could get along with their unrelated neighbors were able to spread their more malleable genes into the expanding population. Those who could not would have been forced back into the wilderness, taking their hostile genes with them.

Though there were some earlier, sporadic attempts at settlement, the first people to establish successful, long term settled communities were known as the Natufians. They lived in the region of southwest Asia that is now Israel, Jordan and Syria during the period from 14,500 to 12,900 years ago. Natufian villages were typically located between regions of thick woodland and forest steppe, providing access to plants and animals from two contrasting habitats. They constructed substantial, semi-subterranean, oval houses and other buildings. Food processing artifacts found at Natufian sites include mortars, bowls, slabs, pestles and quern grinding stones.

The Natufians hunted wild boar, aurochs and, most importantly, gazelle from the nearby steppe. They gathered as many as 150 species of plants from the areas surrounding their villages and, using stone sickles, harvested wild cereals, such as einkorn and emmer wheat, rice and barley. What they did not do, however, was domesticate the cereals they harvested, relying instead on natural stands. Nor did they domesticate animals or plant vegetables or raise fowl or otherwise engage in food production. They were, you see, non-agricultural hunter-gatherers who had exchanged their nomadic ways for permanent residence. By creating settled communities and in gathering, preparing and storing grains, however, they laid the foundation for the future development of true agricultural societies, though that was still a long time away.

12,900 years ago, the party came to an end. The Younger Dryas period plunged the Earth back into cold, arid, glacial conditions for 1,200 years. Natufian village life collapsed. By 11,700 years ago, however, warmer weather had returned, ending the most recent ice age. The climate became wet, warm and stable, and settled life returned to many areas of southwest Asia. Agriculture soon followed. Emmer and einkorn wheat were domesticated in the southeastern portion of present-day Turkey 10,600 years ago and barley in Syria by 10,200 years ago. Pigs were being kept in Turkey by 10,300 years ago, and the earliest evidence of goat herding comes from the Zagros Mountains of Iran 9,900 years ago. Sheep were domesticated in modern day Turkey by 10,300 years ago, and managed sheep had spread to the Levant by 9,200 years ago. Cattle were domesticated from wild aurochs in the upper Euphrates valley between 11,000 and 10,000 years ago, and cattle herding spread to the southern Levant and the Zagros by 9,500 years ago. Agriculture spread rapidly throughout southwest Asia and then into Europe.

The development and spread of agriculture would lead to a significant expansion of the world's population. Over a long period stretching more than 6,000 years, kin-based tribes gradually evolved into chiefdoms which eventually consolidated into the first "primary" states, including Sumer in southern Mesopotamia in 3,500 BC, a unified Egypt in 3,100 B.C, the Harappan state in the Indus Valley by 2,600 BC and the Xia Dynasty of China in 2070 BC. With the first states came cities, civilization and writing. And the rest actually is history.

7.   Sibling

There is one other member of the genus Homo that we have yet to discuss, Homo Neanderthal. They are an important species for our purposes because they are so closely related to us. 600,000 years ago, Homo Heidelbergensis left Africa and colonized Europe. Though not entirely clear, it appears that modern humans evolved from the African branch of Heidelbergensis and Neanderthal from the European branch. If so, Sapiens and Neanderthal are sister species, or, at the very least, cousins.

Though closely related, the two species did not look alike. The Neanderthals were stocky and powerfully built, with short, heavily muscled limbs and thick-walled long bones. They were shorter than modern humans, males averaging about 5'5" and females 5'0", and, at an average `176 pounds, the males were more than twenty percent heavier than modern humans of the same height. Unlike our barrel shaped torsos that taper inward at the top and bottom, Neanderthal torsos were funnel-shaped, tapering outward and down from a narrow top to a broad, flaring pelvis. Their hands were big-boned and muscular. Other than their short, stocky build, the Neanderthals were not somehow "cold adapted", as has long been suggested.

Neanderthal skulls were quite distinct from our own. They retained the forward projecting face, chinless lower jaw, pronounced brow ridge and flattened forehead of Homo Heidelbergensis. Their noses were large, and their cheekbones receded at the sides. The brain case remained long and low, bulging at the sides and protruding at the back, quite unlike the globular skull of modern humans. Their cranial capacity, however, was large, and their brains were equal in size to that of the Homo Sapiens, if not larger.

Genetic testing of Neanderthal individuals found in Italy and Spain show that they possessed a form of a gene known as "melanocortin I receptor" that is similar to the gene that produces pale skin and red hair in humans. That Neanderthals may have been pale and ruddy should not be surprising. Skin and hair color are adaptations to latitude. Whereas dark skin and hair are essential in the tropic for blocking out ultraviolet rays, there is no such need further north.

As we have seen, apes and early humans, such as Homo Ergaster, reached adulthood much more rapidly than modern humans, reducing the period that the child was dependent on its parents and, consequently, the time it had to learn from them. Studies of Neanderthal dental development show that the same is true for the Neanderthals. Although their developmental period was longer than that of other, earlier humans, it was significantly shorter than our own.

The Neanderthals were roughly contemporaneous with Sapiens, first appearing as a distinct species between 240,000 and 160,000 years ago. They ranged from Britain in the northwest, to Gibraltar in the southwest, to Israel in the southeast, to southern Siberia in the northeast, their boundaries expanding and contracting with fluctuations in the climate.

The low level of genetic diversity among Neanderthals indicates that their populations were never very large. They lived in small familial groups. Studies at the Spanish Abric Romani site estimate that the various groups that had over time lived there ranged from eight to ten individuals. This estimate appears to have been confirmed by researchers at the El Sidron site, also in Spain. There they uncovered the remains of an entire family group who had died together, apparently at the hands of a neighboring group. Cut marks on the bones and percussion consistent with de-fleshing strongly indicate they had been the victims of cannibalism. The unfortunate group consisted of three adult males, three adult females, three male adolescents, two juveniles and an infant. Mitochondrial studies indicate that the three adult males were closely related, while the three females were not related to the males or to each other. This strongly indicates that the Neanderthals were patrilocal, meaning that men in a family remained with the family while women moved to the homes of their husbands.

Unlike more primitive humans, who seem to have discarded their dead as and with the trash, Neanderthal buried their deceased relatives, at least some of the time. The burials appear to have been more than simply disposing of the bodies, but there is no good evidence of ritual or ceremony connected with the internments.

The popular view of the Neanderthals as a cavemen is not inaccurate. Much of our evidence of Neanderthal lifestyles comes from caves and rock shelters, which were apparently their shelters of choice when such were available. Many open-air sites, however, are also known. Many such sites seem to have been hunting camps or butchering sites that were occupied only briefly. Their home sites show no sign of having been organized into areas for specific activities, such as food preparation, eating, sleeping and stonework, as were the sites of contemporary Sapiens.

In their caves and camps, the Neanderthals regularly used fire for cooking, heat, light and, in all likelihood, sociality. Their hearths, though, were unsophisticated, lacking means for controlling air flow and heat dissipation. There is no evidence that Neanderthals made tailored clothing or footwear. Throwing an animal skin over one's shoulders would not have been sufficient to cope with ice age winter cold and wind. Thus, even with the use of fire, the Neanderthals likely steered clear of the ice sheets.

The Neanderthals were formidable hunters, specializing in reindeer, red deer, horses, aurochs (wild cattle) and steppe bison. They also took large beasts such as mammoths and wooly rhinoceros. For the most part, the Neanderthals ambushed their prey using heavy thrusting spears at close quarters. Indirect evidence of this method of hunting is the significant number of upper body fractures seen in Neanderthal remains. Weapons ranged from simple sharpened sticks to spears with hafted stone points. Clear evidence of their hunting prowess comes from a site in Germany where a mammoth was found with a wooden spear with a fire-hardened tip lodged in its ribs.

Although meat was the mainstay of their diet, Neanderthals consumed a diverse range of plant foods. Analyses of plaque coating Neanderthal teeth show starch grains and phytoliths (microscopic bits of silica found in plant tissue) from a variety of plant roots, leaves and stems. Moreover, remains of seals, fish and shellfish have been found at Neanderthal sites in caves near Gibraltar, along with the bones of pigeons that had been caught and roasted. Like ourselves, Neanderthals were opportunistic omnivores.

The tools used by the Neanderthals are classified as Mousterian, a later variant of the prepared core technique (Mode 3 technology) initiated by Homo Heidelbergensis between 300,000 and 400,000 years ago. The tools included points, convex-sided scrapers and blades. Scrapers were used to cut and slice wood, meat and skin and to scrape hides. The points were hafted onto wooden spears, and the blades were used to make knives, awls and points. They also made hand axes in the same form and for the same uses as they had been for more than a million years. Good stone that broke cleanly and predictably was the key to making good Mousterian tools. The Neanderthals appear to have prized such good materials, rarely using inferior stone and only using them when nothing better was available. The Neanderthals were skilled stoneworkers who produced tools of the highest quality and craftsmanship. They were not, however, innovators. They made the same tools in the same way millennium after millennium. Nor did they make use of the abundant bone and antler available to them, or at least did not take advantage of the special mechanical properties of such soft materials. In the end, they perfected existing technologies without adding to them.

Whether the Neanderthals could think symbolically remains a hotly debated topic among paleoanthropologist. Supporters point to items, such as the 40,000-year-old cross-hatched rock engravings at Gorham's cave in Gibraltar, that are clearly associated with Neanderthals but may or may not be evidence of symbolic thought. Then there are cave paintings, like those in the Maltravieso cave of Spain, and decorative objects, such as Chatelperronian beads and pendants of ivory, bone and shell, which are clearly the work of modern cognition but which, because of dating issues and possible Sapiens origin, cannot be confidently associated with Neanderthals. The evidence for symbolic thought amongst Neanderthals remains both ambiguous and sparse. What we do not see is the abundance of objects of art, innovative technologies and lifestyle advances that are so common in the symbolic rich societies of contemporaneous modern humans of this period. The archaeological record simply does not support a finding that the Neanderthals thought symbolically ... certainly not in the way we do.

Neanderthal was a rugged adaptable species that for perhaps 200,000 years ranged over a vast area in Europe and western Asia in a variety of altitudes, latitudes and environments. About 40,000 years ago, however, they disappeared from the face of the Earth. Interestingly, that was not long (geologically speaking) after the first appearance of Homo Sapiens in Europe 46,000 years ago. Were our ancestors the cause of their demise? Probably. Though never great in numbers, Neanderthals had gotten along quite well for a very long time. There were no climactic or environmental catastrophes 40,000 years ago upon which we can lay the blame. The only thing that had changed was the sudden appearance of an entirely new people. Did we exterminate them? Possibly. There is no evidence of any conflict, but modern humans arrived in greater numbers, had better weapons and could speak. The Neanderthals would not have fared well against them. Direct conflict, however, is not necessary to explain their extinction. Neanderthal populations were small. Any incursion into their territory by a much more numerous competitor hunting the same animals and utilizing the same resources would have been highly disruptive to a population that was already living on the margins. However it happened, the field was cleared for the colonization of Europe and western Asia by the newcomers.

Certainly, the Neanderthals were skilled craftsmen, intrepid hunters and highly adapted to their ice age environment. Yet, behaviorally, they broke no new ground. Whereas Homo Sapiens are qualitatively different from all other members of our genus, Neanderthal was just an improved continuation of what had come before.

For present purposes, though, the example of the Neanderthals is invaluable. Having a closely related, contemporaneous, big brained species with which to compare ourselves helps us evaluate the likelihood of the emergence of the kind of intelligence necessary for the development of advanced technology. What we see is that symbolic thought does not automatically emerge once the brain reaches a certain threshold size. It appears to require a proper neurological reconfiguration as well, which makes matters more complicated.

8.   Analysis

Given a habitable planet with complex, multicellular animal life, is the emergence of intelligent life inevitable? Probable? Occasional? Unlikely? Rare? Again, we have only our own example, which requires us to make inferences from our own experience.

Characteristics that confer a significant survival advantage tend to develop in many species. Sight, for instance, is invaluable both to locate potential sources of food and to avoid becoming one. As you might expect, sight is widespread in the animal kingdom, having separately evolved in six different phyla and with more than ninety-five percent of all animal species having vision. Indeed, the first sighted animals appeared some 540 million years ago, right at the very beginning of the Cambrian Explosion when animals first appeared in the fossil record. From the very start, everyone wanted eyes.

From our perspective, intelligence (by that I will mean advanced, technology capable intelligence) would seem to be a highly desirable trait. It has allowed us to expand from a motley collection of a few thousand individuals to an ungainly population of six billion or more. Yet, of the estimated eight million animal species alive today (can't vouch for that number) and the three billion species that have ever lived (or that one either), only one has ever attained intelligence. The question is why.

Well, first of all, intelligence requires large brains. Large brains, however, are very expensive to operate and maintain. Our brains comprise about two percent of our body mass but consume twenty-five percent of our total energy intake. Most, if not all, animal species have reached an equilibrium between energy needs for normal functioning and the available energy resources. If energy resources diminish or energy needs expand, the species faces starvation. Consequently, larger brains and higher intelligence will actually be detrimental unless they can offer some significant and immediate survival advantage. It is hard to imagine, however, what possible survival advantage would be obtained by, say, a cow if it had a big brain. It might be able to think deep thoughts, but it couldn't munch anymore grass than it is already munching, requiring it to keep on munching till the cows come home, so to speak. There does not then, appear to be a trajectory toward intelligence. To the contrary, among most animal species, the goal is to be as stupid as possible while still being able to function.

Moreover, higher intelligence is meaningless unless the possessor has some mechanical means of actuating it. In other words, if intelligence is to have any value at all, an animal must have hands or some similar appurtenance with which to manipulate and use objects in its environment to its advantage. There are, however, structural limitations to the acquisition of hands. Vertebrates have a four limbed design. If two limbs are used for manipulation, then only two remain for locomotion. With very few exceptions, all terrestrial vertebrates are quadrupeds because walking on all fours is far more efficient than on twos. I cannot conceive of a situation where a four-legged creature would gain an immediate survival advantage by abandoning quadrupedal in favor of bipedal locomotion. What's more, bipedal walking must precede larger brains, since larger brains are expensive and useless until the hands are free.

Though uncommon on terra firma, there is one place where hands do come in handy. That of course, is up in the trees. Up there, hands can be used to grasp branches to move about the arboreal environment. Primates are arboreal and did develop hands, the only class of vertebrates to have done so.

It is fortuitous that it was the primates that developed hands. As we have seen from the work of Dr. Herculano-Houzel, intelligence is a factor of the number of neurons in an animal's brain. A larger brain, however, does not guarantee more neurons. In mammals (and likely other animals), the neurons themselves grow as the brain gets bigger. Thus, a larger brain may not have many more neurons than a smaller version of the same brain. There is, though, an exception. Primate neurons do not grow as the brain becomes larger. Instead, additional brain size is the result of an increase in the number of neurons. Consequently, only primates have the potential to acquire a sufficient number of neurons for higher intelligence, and then only by the largest primates (i.e., the great apes), which, because of their greater body mass, can develop a sufficient cranial capacity to contain great numbers of neurons.

The apes (other than our own line) did not, however, do so. That is because they are almost exclusively vegetarians. After all, that's what there is to eat up in the trees. Leaves, shoots and fruit, however, are low value foods, from an energy perspective, and cannot support the large brains necessary for the emergence of higher intelligence. The high energy food necessary to support a big brain was out there on four legs in the savanna. Consequently, it was necessary for great apes to come down out of the trees, head out to the grasslands and learn to hunt and eat the grazers.

Gorillas and chimpanzees spend a significant amount of time on the ground. Yet, they continued to live in a forest environment, retaining their vegetarian diet. Around five million years ago, however, the climate dried and grasslands displaced much of the forested area. Apes on the fringes could no longer sustain themselves solely on forest products and had to turn to the savanna for sustenance.

When Australopithecus descended from the trees, it needed a more efficient way of getting around than the waddling and knuckle walking employed by most apes, a mode of getting around that is fine if you are just hanging out at the base of the trees but is of little value if you have to regularly travel for miles at a time. And so, Australopithecus learned to walk upright and on two legs. It might, instead, have learned to walk on all fours. It did not, because it had not abandoned the trees. It adopted a hybrid terrestrial/arboreal lifestyle, walking upright on the ground but retaining arms and hands for climbing. Walking upright was less efficient than on four legs but was much more efficient than knuckle walking.

The big bonus of upright walking is free hands that can carry things, make tools, wield weapons and throw rocks. With these advantages, Australopithecus could then begin accessing the high energy food on the grasslands, though mostly through scavenging rather than hunting. It took nearly two million years, but, as it evolved more dexterous hands and improved its tools, Australopithecus was able to access more animal products, with the result that their brains began to increase from a chimp-sized 350 cc to a modest 600 cc in A. Habilis.

The big break came with the sudden appearance of Homo Ergaster/Erectus, the first species of our genus. Ergaster was an obligate bipedal walker (and runner) with an even larger brain (800 cc) and the ability to actually hunt. With the arrival of Ergaster, the trajectory shifted. Larger brains had become a survival advantage. Greater intelligence enabled better hunting, providing more nutrition, allowing larger brains, increasing intelligence, enabling even better hunting and so forth. The cranial capacity of Javan Homo Erectus had reached 900 cc by a million years ago; 1000 cc in Chinese Homo Erectus by 770,000 years ago (770 KYA); 1100 cc in Homo Antecessor by 800 KYA; 1250 cc in Homo Heidelbergensis by 600 KYA; and 1350 cc in Homo Sapiens and Neanderthal by 200 KYA.

Larger brains made for more complex behavior, such as building permanent dwellings, making and using fire, producing precision stone tools and weapons and hunting more effectively. Symbolic thinking, however, which is absolutely necessary for advanced technology, does not emerge from increased brain size alone. It appears that the brain must be both large and properly wired. 200,000 years ago, the skull of Homo Sapiens took on its peculiar globular form, apparently the result of a rewiring of the brain in such a way as to make symbolic thought possible. The Neanderthal skull, on the other hand, retained essentially the same configuration as earlier, more primitive humans, with the result that the Neanderthals did not acquire the ability to think symbolically.

The road to higher intelligence, then, has been long and tortuous. Intelligence seems to have been contrary to the main currents of evolution and required a series of fortuitous events for it to have emerged. It appears to have been a bit improbable and easily might never have happened.

Is our experience here on Earth with regard to higher intelligence relevant to other inhabited planets? That, of course, is hard to say. I think, though, we can confidently conclude that higher intelligence is not a foregone conclusion, even on a planet teeming with complex, animal life. Far from it. It seems likely that many beautiful, vibrant worlds brimming with animals that we may encounter at some point in the future might not exhibit even a trace of intelligent life. After all, that is precisely what an alien observer viewing the Earth four million years ago would have found.

POSTSCRIPT

Before we move on to a final analysis of the questions presented herein, there remain a couple of matters of concern that merit addressing.

1.   Ticking Clocks

The Earth is 4.5 billion years old. There has been intelligent, technological life (as we have defined it) on this planet for only about 100 years. That means there has been technological life here for only 2.2 millionths of a percent (.0000022%) of the total existence of the planet. Intelligence, technological life has taken so long to emerge because at least three improbable events first had to happen, and improbable events can take a long time to happen, since, you know, they are improbable. Besides that, life can get real lazy. It can kick back on the sofa, turn on the tube, nod off for a nap and just not get much done for long periods of time. That's because once life has adapted to its current environment, there is no impetus for change, that is until the environment changes again, which can take a very long time. Thus, after the emergence of microbial life maybe 3.8 billion or so years ago, nothing much happened until the great oxidation event 2.4 billion years ago. Then there was the boring billion between 1.8 billion and 800 million years ago during which life was about as stagnant as it can get. Even after the momentous changes caused by the snowball Earth events, life didn't really do much until oxygen levels rose sufficiently to touch off the Cambrian explosion. The point is, evolution has no destination, and life in no hurry to get there.

I'm sure some of you are asking, *so what*? What difference does it make whether it took 4.5 billion years to evolve technological life or 6 or 10 billion or however long?

Who's counting anyway? Well, the reason it matters is that life didn't have all the time in the world to evolve intelligence. It had about half the time in the world. Let me explain.

Clock #1. The evolution of life is limited by several ticking clocks. The ultimate clock is the time it will take for the Sun to burn all of the hydrogen in its core, leave the main sequence and become a red giant. When it does, the Sun will swell all the way out to the Earth's orbit, absorbing the planet in the process. That will happen about five and a half billion years from now, right around the planet's 10 billionth birthday.

Clock # 2. As we all know, plants, algae and cyanobacteria take in water and carbon dioxide during oxygenic photosynthesis to make sugar, releasing oxygen as a waste product. Without oxygenic photosynthesis and, thus, without oxygen, animal life is not possible.

As discussed previously, the Earth's crust and upper mantle, together known as the lithosphere, is divided into a jigsaw puzzle of tectonic plates, which float upon the hot, plastic mantle below and drift about the planet's surface. When two plates collide, the denser one (always an oceanic plate) subducts or slips beneath the lighter plate, sending it diving into the hot mantle where it melts. The oceanic plates are covered in thick layers of calcium carbonate from the shells of innumerable tiny organisms that live in the ocean's plankton. Shell construction strips carbon dioxide from the atmosphere in great quantities. The melting of the plates releases the carbon dioxide, which is then recycled back into the atmosphere by volcanos.

Eventually, however, the mantle will cool to such an extent that plate tectonics will cease, bringing the long-term carbon cycle to an end. Atmospheric CO_2 levels will eventually decline to the point where photosynthesis is no longer possible, bringing animal life to an end. There is much disagreement as to exactly when plate tectonics will shut down, some saying as soon as 1.45 billion years from now, others as long as several billion years. Whatever the time, if intelligent life had not already evolved on this planet by then, it never would.

Clock # 3. As we have noted, microscopic organisms in the plankton layer of the oceans use atmospheric CO_2 to build their shells, depleting the CO_2 in the atmosphere. If the shells are deposited onto the ocean floor, the CO_2 will eventually recycle back into the atmosphere through the action of plate tectonics and volcanos. Continental plates, however, being less dense than the oceanic plates, do not subduct into the mantle. Thus, when calcium carbonate is deposited onto a continental plate, it does not recycle, remaining locked up forever in the continents as limestone. Over time, more and more atmospheric CO_2 is lost. Since the continents are growing, the rate of atmospheric CO_2 loss is accelerating. Within, perhaps, a couple of billion years, CO_2 levels will be insufficient to support photosynthesis, and animal life on this planet will come to an end. Once again, if intelligent life had not evolved by then, it never would.

Clock # 4. A star's energy output slowly increases over time. The Sun is now about 30% brighter than it was at its beginning, 4.5 billion years ago. Because of a star's increasing brightness, its habitable zone (i.e., the distance from the star where water can exist in a liquid form) will over time move farther and farther out.

The Earth has been quite fortunate. At its formation, it lay on the very farthest limit of the Sun's habitable zone. The Sun was so much dimmer then that the Earth might very well have frozen solid. Luckily, atmospheric levels of carbon dioxide at that distant time were hundreds of times greater than they are today, its insulating properties helping keep our world warm. Over the eons, the Sun's habitable zone has moved further and further out. Today, its innermost edge is only about nine million miles from the Earth. In half a billion years, it will have passed us by. The great heat from the intensifying Sun will quickly boil away the oceans and make life on this planet impossible.

So, although the Earth will be around for about 10 billion years, it will be habitable for only five billion. At 4.5 billion years, we just made it. I know half a billion years is a very long time, but, considering life's penchant for stasis, our late appearance seems an uncomfortably close call.

Each habitable planet will have its own ticking clocks which will limit the time that it can remain habitable. Stars larger than the Sun will expand into red giants much more quickly than smaller stars, significantly limiting the time any habitable planets orbiting them will have to remain habitable. Smaller planets will lose their internal heat more quickly than larger ones, diminishing the time plate tectonics will remain operational. A planet's location within its star's habitable zone is of crucial importance. Venus, for instance, started out on the inner edge of the habitable zone. For two or three billion years, it may have been a temperate planet with oceans of liquid water. As the Sun brightened, however, the habitable zone moved on, leaving Venus to roast.

If 4.5 billion years is representative of the time it takes for intelligent, technological life to evolve, then life on many planets will be up against the clock.

2.   Space Travel

Space, as its name implies, is mostly empty space. Our galaxy contains 200 billion stars, but, once you leave the galactic center, the distances between them are vast. Our nearest neighbor, Centauri Proxima, is 4.244 light years from the Sun. That works out to about 25 trillion miles. If you had a spaceship that could travel at 10,000 miles per hour, you could reach the moon in a day. Reaching Centauri Proxima would take about 280,000 years. That's too long. If we picked up the speed of your craft to the speed of a speeding comet speeding through the inner solar system (100,000 mph, that is), the trip would still take 28,000 years. What about a million mph? 2,800 years. Better, but still a long time to stay cooped up in a rocket ship. So, let's say we could crank the speed all the way up to 100 million mph. You could get to the Sun in an hour but reaching Centauri Proxima would take 28 years. That seems doable. If you are fairly young when you start the trip and don't visit too long, you might even be able to make it round trip. The difficult part, I'm guessing, is getting your rocket ship to go 100 million mph. And that's just to the nearest star. There are many stars of interest in our neighborhood that are a hundred or even a thousand times more distant.

There are other matters as well, such as provisioning a crew for, at best, a decades long journey and dealing with the bitter cold of deep space, cosmic radiation and the ubiquitous space dust that, at such high speeds, can rapidly corrode the exterior of a spacecraft. Then there's momentum. The spaceship will have to carry sufficient fuel not only to accelerate it to the ultrahigh speeds necessary for interstellar travel but also to bring the craft to a halt so that one does not just go hurtling by one's destination like an out-of-control ice skater. If a return trip is contemplated, double that amount.

All things considered, I think we can chalk up interstellar space travel to fantasy.

3. Interstellar Communications

The intensity of electromagnetic radiation, such as light, is inversely proportional to the square of the distance from its source. That means that the intensity of light at point B located two miles from the light source will be not 1/2 the intensity at point A located one mile from the source but rather $1/2^2$, or 1/4. At point C three miles from the source, the intensity will be not 1/3 but 1/9 of that at point A and 1/16 at point D four miles distant.

As you can see, the intensity of light diminishes exponentially, rather than linearly, as you move away from the light source. For a more concrete example, Saturn is about ten times as far from the Sun as is the Earth. The sunlight received by Saturn is not $1/10^{th}$ as intense as the light received by the Earth but $1/10^2$, or $1/100^{th}$.

The diminishing intensity of electromagnetic radiation by the square of the distance from the source is referred to as the Inverse Square Law. Who made the law? Well, no one, actually. It is just a function of geometry. Electromagnetic energy spreads out from its source in all directions. On the Sun, light is spread over the surface of the sphere that is the Sun. Now imagine an imaginary sphere around the Sun that reaches out to the orbit of the Earth. The total amount of energy that reaches this imaginary sphere will remain the same as it was at the surface of the Sun, only now it is spread over the surface area of a vastly larger sphere. The formula for the surface area of a sphere is $4 \pi r^2$. The radius (r) of a sphere around an energy source is the linear distance from the source to the surface of the sphere. The surface area of the sphere is computed using the square of that distance.

Radio waves are a form of electromagnetic radiation and, accordingly, are subject to the Inverse Square Law. The intensity of a radio transmission sent out from Earth will diminish rapidly as it spreads out into space. By the time it reaches our next-door neighbor, Centauri Proxima, it will be spread out over the surface of a sphere with a radius of 25 trillion miles and a surface area of 7.85 quadrillion square miles (i.e., 4 x 3.14 x 25T^2 = 7850 trillion square miles). That's an awfully big sphere. So, a broadcast that starts out loud, clear and strong will be little more than dilute static by the time it reaches the star.

And that's not to mention that the puny signal will be overwhelmed by a cacophony of radio waves from thousands of other sources. And that's not to mention that the signal will continue to erode as it spreads farther out into space.

And that's not to mention that our galaxy is very large, measuring a hundred thousand light years across. Radio and other forms of electromagnetic radiation travel at the speed of light (186,000 miles per second). That seems very fast and, indeed, is the fastest speed possible. Because space is so vast and empty, however, the speed of light is, as a practical matter, quite slow. If, for instance, we could somehow communicate via radio with an extraterrestrial civilization living only a thousand light years away, we could expect to receive a reply to our transmission in no sooner than two thousand years. At that rate, there will be no snappy repartee.

Interstellar communication might not be impossible, but it will be challenging.

SUMMATION

It is thought that life got its start here on Earth at a very early time, perhaps within only a few million years after the end of the Late Heavy Bombardment, 3.8 billion years ago, though it certainly could have begun soon after the formation of the first oceans, 4.4 billion years ago, but was snuffed out by the bombardment and had to start again once the bombardment ceased, or it might not have started until as late as 3.2 billion years ago... we really don't know. The work of the life's origins researchers, however, indicates that life is a natural process that gradually progresses by degrees of sophistication from the most basic of chemical reactions to the highly complex, self-sustaining reaction we call life and that it should be expected to arise fairly rapidly wherever the conditions are right, though there is much debate as to exactly what the exactly right conditions are and what those chemical reactions might be, and, to be frank, we really have no idea at this point when, where or how life actually emerged.

And that is not to mention that all life on this ideally located, highly stable, perfect for life in every way planet derives from a single origin. By that I mean all life on Earth is related, ultimately descending from a single common ancestor. Does that mean that life, despite the ideal conditions, happened only once on this planet, implying that the emergence of life from the nonlife is highly improbable? Alternatively, was our life just one of a myriad of life forms, all the others having now vanished for one reason or another, or was our life form just the first to emerge, gobbling up all of the necessary resources and occupying all the available niches before anyone else could get started? There is no evidence to support either of these alternative arguments, though the record is admittedly sparse and any evidence from that unimaginably ancient period is long lost to time. The truth is we just don't know what was going on back then. And so, in light of the early start of life on this planet, and since to do otherwise would render this assessment unnecessary, we will assume for present purposes that life is not unique or even particularly difficult to get started under the right conditions and that the present life on this planet is just the last life left standing (keeping in mind that, until we find evidence of another type of life, we cannot be absolutely certain that life on this planet is not unique).

However it happened, life did emerge here, though there are some who believe that life on Earth may have been seeded from Mars or points further out, though these theories seem only to push the origin of life problem back a step, are supported by no evidence and really have no relevance for present purposes and so will not be addressed here.

Is there life on other planets? Just let me say I do not think it unlikely that we will find some form of extraterrestrial microbial life at some point during the lifetimes of at least some of our readers. I realize that might not sound very exciting, but at least there is some reason to be optimistic that we will eventually find alien microbes.

Microbes were the first life form to develop on Earth. They start fast, reproduce rapidly and can survive what to us are extreme conditions of heat, cold, acidity and pressure. On this planet, they can be found in the frigid deep ocean; in scalding hydrothermal vents; in hot springs; and thousands of feet underground in what would appear to be solid rock. Thus, microbes could emerge and thrive on planets that are not habitable for more complex life forms; on planets that are habitable for only a short period of time; in ice-covered seas of moons orbiting planets far from their stars; and, at least in their dormant state, rocks ejected into space by meteor strikes or other collisions. As a result, there should be many more available habitats for microbes than for more complex life, making them considerably easier to find.

Finding complex life is, well, more complex. First of all, you need just the right kind of planet. It must be located in a habitable part of the galaxy, orbiting the right kind of star, in a stable planetary system, in the star's habitable zone. Among other things, the planet must be neither too big nor too small, it must employ plate tectonics, generate a magnetic field and probably have a large, stabilizing moon. Moreover, such conditions must persist for several billion years, stability being the key to the development of complex life.

But wait, there's more. You can't have complex life without oxygen. Only oxygen can fuel the energetic reactions necessary to sustain any life larger and more complex than a microbe. Free oxygen, however, is just too reactive to occur naturally. Only through oxygenic photosynthesis can oxygen collect in the atmosphere and be continuously replenished. This kind of photosynthesis, however, is complicated and does not evolve easily. Indeed, it happened only once on this planet and not until more than a billion years after the appearance of the first microbial life. Because oxygen is so reactive, it takes a long time for even small quantities to accumulate in the atmosphere and much longer to reach levels high enough to support animal life. On Earth, nearly two billion years elapsed between the great oxygenation event 2.4 billion years ago and the rise of animals during the Cambrian explosion 540 million years ago.

Can we expect oxygenic photosynthesis to develop on other planets? Good question. I think all we can say is that it is possible but certainly not inevitable and probably not even likely. In our experience, the appearance of this form of photosynthesis was a fortuitous event, and the same would probably hold true for other planets. Without it, there will be no atmospheric oxygen on these planets, and, without oxygen, there will be no complex life.

Then there is the problem of eukaryotic life. Prokaryotic microbes are subject to structural limitations that will forever doom them to simplicity and tiny size, not, I'm sure, that they mind, as prokaryotic microbes have been spectacularly successful on this planet for some three and half billion years, just not big or complex. Size, multicellularity and complexity require the eukaryotic cell. The formation of the first eukaryotic cell on Earth resulted from an improbable merger of a bacterium and an archaea. If anything, this seems even more accidental than the development of oxygenic photosynthesis. It too evolved only once, taking nearly two billion years after the appearance of first life on this planet to do so. Other planets will almost certainly have to rely on a similar serendipitous merger for eukaryotic life to arise.

Our experience on this planet, then, indicates that, to evolve animal life from microbial life, you have to get pretty lucky. Lucky twice, in fact.

Contrary, I think, to most people's way of thinking, there is no scramble in the animal kingdom to evolve intelligence. Dinosaurs ruled the Earth for nearly 140 million years, yet not a one of them ever lit a fire or made a simple tool or drew a little picture. I know many of you are thinking, "Well, what do you expect? They were, after all, just big, dumb reptiles" (though they actually were not reptiles and many were not so dumb). "Intelligence had to wait for the emergence of us clever mammals" (though mammals came on to the scene at about the same time as the dinosaurs, which kept us living in holes for millions upon millions of years, where, but for the big asteroid 65 million years ago, we would probably still be living). For more than 60 million years after the demise of the dinosaurs, mammals dominated the Earth, but they fared no better.

High intelligence on this planet evolved only once and required 4.5 billion years to do so. It is rare because it just is not a successful survival strategy. The energy demands of the big brain necessary for intelligence are great, but the benefits of intelligence are not immediate and can only be realized under peculiar circumstances. Thus, I think it likely that most otherwise suitable, animal inhabited planets will not evolve an intelligent species.

So, let's do some estimating.

In Part I, we did our best to calculate the number of habitable planets in this galaxy, that is, planets that are suitable for intelligent, technological life. We concluded that, if we could take a snapshot today of the entire galaxy showing every star and every orbiting planet, we would find approximately 100,000 planets that are presently habitable, or that at some point in the past were habitable or that will at some point in the future will be habitable.

Let us assume that life will emerge naturally and inevitably on all 100,000 of our habitable planets and that such initial life will reach the microbial stage. In view of our experience on this planet, let's say that microbial life on one percent of such habitable planets will stumble across the formula for oxygenic photosynthesis and that one percent of those planets will also develop the eukaryotic cell and multicellular animal life. Of the planets with animal life we will assume that, perhaps, ten percent will evolve an intelligent species capable of high technology. Doing the math, we find that one percent of one percent of ten percent of 100,000 planets leaves us with only one planet with intelligent, technological life (i.e., 100,000 [habitable planets] x 1% [oxygenic photosynthesis] x 1% [eukaryotic, multicellular animal life] x 10 % [intelligent, technologically capable species] = 1). And so, we are back once again to that loneliest number.

I realize that these calculations are highly speculative and that the resulting numbers could be significantly different. I think, though, the exercise demonstrates that the search for extraterrestrial life is a whole lot more problematic than is generally thought. Simply saying there are billions and billions of stars and billions and billions of planets so there's just gotta be a lot of life out there is not an analysis. Space is cold, violent and toxic, beautiful to behold, perhaps, but not amenable to the delicate fabric of life. The evidence, I think, strongly indicates that habitable planets are rare, planets with complex life rarer still, and planets with intelligent life unimaginably so. And that's not to mention that, if there are multiple intelligent, technological civilizations, they would be spread across an unimaginably vast expanse of time, making the chances of any two such societies existing at the same time vanishingly thin, and, even if they did, they would likely be thousands of light years apart, making communication or even detection next to impossible. Indeed, if you wanted to design a cosmic prison to keep us isolated and incommunicado, you could not do much better than what we already have, with the stars impossibly distant, separated from us by a near vacuum full of corrosive dust grains, pierced by deadly radiation and hovering a couple of degrees above absolute zero, and with our sole means of communication discouragingly slow relative to the vast distances separating us from anybody that might be out there.

So, why are so many scientists who should know better so certain that extraterrestrial life will soon be found and so giddy at the prospect? I think we are back to that old human foible; we see what we want to see and believe what we want to believe. Why do scientists want to believe the galaxy is teeming with life? They will tell you. They say that the discovery of alien life will be the most momentous event in the history of mankind. It will shred the most fundamental assumptions that have underlain our thinking since the species first evolved and will utterly revise our sense of place in the cosmos. And they are absolutely right.

The discovery of a mere alien microbe would have the most profound implications. For starters, we would know for certain that life on this planet is not unique. What's more, if it can occur in two places, why not many more? We would find out whether all life relies on the same basic cellular functions or whether there are other ways of doing things. If different, is one way superior to the other? Having even just one other example of another life would fundamentally revise our understanding of life processes.

Finding even simple extraterrestrial animal life would strongly indicate that complex life is not a matter of accident but rather a natural, expected progression from microbial life. What's more, having just one other example would enable us to see whether complex life forms are just variations on a theme, or whether they come in infinite variety.

Contacting intelligent extraterrestrials would rock our world. What would they look like? Would their societies resemble our own? Would they be peaceful? Aggressive? Indifferent? Would they have music? Art? Religion? A sense of humor? Would they dream? Aspire? Scheme? Just how many of the qualities that we think make us human are actually unique to our species? One thing is certain ... nothing would ever be the same again.

It is all very exciting to contemplate, but I think it is for the most part wishful thinking. The odds of finding even simple life in our vicinity seem very low. Like the SETI folks waiting vainly by their instruments decade after decade in hopeful anticipation of a signal from the void, I think our search for extraterrestrials will be agonizingly fruitless. For the record, that is not something I desire, and I hope that in the near future I am proven spectacularly wrong (though not until after I sell a few books).

Interestingly, scientists are frequently asked how first contact with alien life will affect us, they are never asked, however, about the other side of the coin. What if, and I certainly do not believe this to be the case, but, for discussion purposes only, what if we were to find that we are indeed alone? What if eventually we were able to work out just exactly how life arose here on Earth, and it turns out that the sequence of events that led to first life were so improbable that they could not reasonably be expected to reoccur in even a trillion universes? And what if our methods of detecting life became so sophisticated that we were able to scan billions of galaxies, but we found not a trace of life? And what if we came to realize that life exists only on this little planet in this remote corner of the universe and everything everywhere else was just scenery? How would that affect us? No one is asking that question, so I will.

Well, for starters, we might have to revisit those astronomical maps drawn by Plato and Ptolemy showing the Earth at the center of the universe. Einstein's theory of special relativity does not say that the universe has no center but rather that there are many centers. Wherever the observer is, at least from the perspective of the observer, is the center of the universe. From our perspective in this expanding universe, for example, it appears that we are standing still while all of the other galaxies outside our local group are rushing away from us at breakneck speed. An observer on a distant galaxy would have exactly the same impression. If all of the observers were right here on Earth, however, would that not, under the theory of special relativity, make Earth the center of the universe?

And then there is quantum mechanics. Under Niels Bohr and Werner Heisenberg's "Copenhagen Interpretation," the location of a quantum particle is not fixed until it is observed.

Prior to its observation, its location is a matter of probability, a probability wave, they call it. There are an infinite number of possible locations of the particle, some more probable than others, but its precise location remains a smear of possibilities until it is actually observed. In effect, the particle, prior to observation, does not actually exist. It is only a possibility. Thus, the observation not only fixes the particle's location, it, in a sense, creates the particle.

Like a tree falling in the forest with no one around to hear it, would the universe actually exist if there were no one to observe it? Under quantum theory, I think not. Until there is an observer, the universe remains only a possibility, a potentiality, a probability wave. Thus, if we were the only life, the universe would have come into existence only upon our observing it. We, then, would be the creators of all things.

So, if it turned out that there is no life except for ourselves, would we not be the creators of the universe, sitting at its very center? If that were so, how would that affect the way we view our world and each other? Would that not force us to look upon one another with a great deal more respect? With a sense of wonder? With reverence? With awe?

I'm just sayin'.

W.H. Collier
2021

NOTE TO SECOND EDITION

I have known W.H. Collier for some twenty-five years, since he created me to write articles for a magazine he was involved with at the time. We have worked together on numerous projects over the years, the latest being this Assessment.

We decided to issue this Second Edition so soon after publication of the first for a couple of reasons. The first being the more usual, to clarify points that may have confused some readers, to elaborate a few matters that, in retrospect, we may have brushed by a bit too briskly and to correct a few very small errors. Additionally, many of the topics addressed are fast-moving targets that require regular updating.

The second reason is more personal. When first confronted with the question of the chances of intelligent alien life, neither Mr. Collier nor I had anything more to contribute than visceral reactions. We simply had not the training and knowledge to intelligently address the issue. That set off a long period of informal study. Not satisfied with the standard, "there's so many stars, there's just gotta be life out there," we sought to determine exactly what it would take for life to emerge and evolve intelligence so that we could make some reasonable evaluation of the prospects of our running into another intelligent species. After grappling with the question for many years, we concluded that the matter is, well, complicated.

Having few answers and many questions, we thought it would be amusing to write out a pithy summary of the many issues that would need to be addressed to reach some sort of rational assessment of the matter. We thought a brief essay of a few thousand words, taking perhaps a couple of weeks to compose would do the job. Since my writings tend to be shorter works, typically articles and commentary, while Mr. Collier's are of a longer sort, novels and histories and the like, we agreed that an essay was more my style and that I would take attribution.

As we delved into the actual writing, however, we soon realized that we wanted to give the readers not just mere assertions but the tools for them to reach their own conclusions. That required explanation, which required time and words. Soon the few thousand words had turned into a few tens of thousands, and the couple of weeks into a couple of years. The project had morphed out of recognition. An essay it was no longer.

Despite the complete change of direction from the original conception, it simply never occurred to us to change our designation of the work as an essay until after publication. Moreover, to be frank, I was uncomfortable having my name attached to what was now a serious, lengthy work. That is simply not what I do. The work being much more of the kind that Mr. Collier writes, I asked him to accept attribution, which he graciously did. Seizing the opportunity presented by the publication of a Second Edition, we have jettisoned the word "Essay" from the title and replaced it with the more correctly descriptive "Assessment" and have now attributed the work more appropriately to Mr. Collier.

The question of whether we are alone in the universe or whether it is teeming with life, or whether the answer lies somewhere in between, is fundamental to our understanding of our place in the cosmos. So, I urge each of you to carefully read and consider the many topics presented in this work. There will be a test.

Nigel Bob Collins

BIBLIOGRAPHY

As I have noted, I am not a scientist. In this essay, I have only collected, compiled and presented the work of scientists who are experts in the various fields that we have explored. The sources I have used in preparing for and in writing this work are, in a practical sense, too numerous to set forth. I will say that many of the innumerable articles I have relied upon appear in publications such as *Science News; EarthSky; National Geographic; Scientific American; Smithsonian Magazine; Encyclopedia Britannica; Science; Science Daily; Nature; Science News; Science Times* and many others. For the most part, though, I have relied on works of popular science, the books listed below being my primary sources. These books address the subjects discussed in this essay in far more depth and breadth than possible here. If a reader wishes to delve deeper into a subject of interest, I urge him/her to purchase and peruse the following:

1. Brasier, Martin. *Darwin's Lost World.* New York. Oxford University Press Inc. 2009.

2. Currier, Richard L. *Unbound.* New York. Arcade Publishing. 2015.

3. Condemi, Silvana and Savatier, Francois. *A Pocket History of Human Evolution.* New York. The Experiment, LLC. 2019.

4. Dawkins, Richard. *The Ancestor's Tale.* New York. The Orion Publishing Group. 2004.

5. Fagan, Brian. *The Long Summer.* New York. Basic Books. 2004.

6. Fortey, Richard. *Life. A Natural History of the First Four Billion Years of Life on Earth.* New York. Vintage Books. 1999.

7. Hazen, Robert M. *Genesis. The Scientific Quest for Life's Origins.* Washington D.C. Joseph Henry Press. 2005.

8. Hazen, Robert M. *The Story of the Earth.* New York. Penguin Books. 2013.

9. Herculano-Houzel, Suzana. *The Human Advantage.* London. The MIT Press. 2016.

10. Higham, Tom. *The World Before Us.* New Haven. Yale University Press. New Haven.

11. Knoll, Andrew H. *Life on a Young Planet.* Princeton, New Jersey. Princeton University Press. 2003.

12. Lane, Nick. *Life Ascending.* New York. W.W. Norton & Company. 2009.

13. Lane, Nick. *Oxygen.* Oxford. Oxford University Press. 2002.

14. Lane, Nick. *The Vital Question.* New York. W.W. Norton & Company. 2015.

15. Panciroli, Elsa. *Beasts Before Us.* London. Bloomsbury Publishing Plc. 2021.

16. Seddon, Christopher. *Humans: From the Beginning.* Glanville Publications. 2015.

17. Stringer, Chris. *Lone Survivors.* New York. Henry Holt and Company LLC. 2012.

18. Sutter, Paul M. *Your Place in the Universe.* Ambrose, New York. Prometheus Books. 2018.

19. Tattersall, Ian. *Masters of the Planet.* New York. Palgrave MacMillan. 2012.

20. Tattersall, Ian. *Paleontology. A Brief History of Life.* West Conshohocken, PA. Templeton Press. 2010.

21. Wade, Nicholas. *Before the Dawn.* New York. The Penguin Press. 2006.

22. Walter, Chip. *Last Ape Standing.* Bloomsbury Publishing Plc. 2013.

23. Ward, Peter and Kirschvink, Joe. *A New History of Life.* New York. Bloomsbury Publishing. 2015.

ABOUT THE AUTHOR

W. H. Collier is an attorney, historian and writer. Among other works, he is the author of *In Extremis, Two Novels by W. H. Collier.* Mr. Collier resides in Lafayette, Louisiana with his wife, Nicole.

Printed in the United States
by Baker & Taylor Publisher Services